EAI/Springer Innovations in Communication and Computing

Series editor
Imrich Chlamtac, European Alliance for Innovation, Gent, Belgium

Editor's Note

The impact of information technologies is creating a new world yet not fully understood. The extent and speed of economic, life style and social changes already perceived in everyday life is hard to estimate without understanding the technological driving forces behind it. This series presents contributed volumes featuring the latest research and development in the various information engineering technologies that play a key role in this process.

The range of topics, focusing primarily on communications and computing engineering include, but are not limited to, wireless networks; mobile communication; design and learning; gaming; interaction; e-health and pervasive healthcare; energy management; smart grids; internet of things; cognitive radio networks; computation; cloud computing; ubiquitous connectivity, and in mode general smart living, smart cities, Internet of Things and more. The series publishes a combination of expanded papers selected from hosted and sponsored European Alliance for Innovation (EAI) conferences that present cutting edge, global research as well as provide new perspectives on traditional related engineering fields. This content, complemented with open calls for contribution of book titles and individual chapters, together maintain Springer's and EAI's high standards of academic excellence. The audience for the books consists of researchers, industry professionals, advanced level students as well as practitioners in related fields of activity include information and communication specialists, security experts, economists, urban planners, doctors, and in general representatives in all those walks of life affected ad contributing to the information revolution.

About EAI

EAI is a grassroots member organization initiated through cooperation between businesses, public, private and government organizations to address the global challenges of Europe's future competitiveness and link the European Research community with its counterparts around the globe. EAI reaches out to hundreds of thousands of individual subscribers on all continents and collaborates with an institutional member base including Fortune 500 companies, government organizations, and educational institutions, provide a free research and innovation platform.

Through its open free membership model EAI promotes a new research and innovation culture based on collaboration, connectivity and recognition of excellence by community.

More information about this series at http://www.springer.com/series/15427

Huimin Lu • Li Yujie

Editors

2nd EAI International Conference on Robotic Sensor Networks

ROSENET 2018

Springer

Editors
Huimin Lu
Department of Mechanical and Control
Engineering
Kyushu Institute of Technology
Kitakyushu, Japan

Li Yujie
Department of Electronics Engineering
and Computer Science
Fukuoka University
Fukuoka, Japan

ISSN 2522-8595   ISSN 2522-8609   (electronic)
EAI/Springer Innovations in Communication and Computing
ISBN 978-3-030-17762-1   ISBN 978-3-030-17763-8   (eBook)
https://doi.org/10.1007/978-3-030-17763-8

This Springer imprint is published by the registered company Springer Nature Switzerland AG.
The registered company address is: Gewerbestrasse 11, 6330 Cham, Switzerland

Preface

We are delighted to introduce the proceedings of the 2017 European Alliance for Innovation (EAI) International Conference on Robotic Sensor Networks (ROSENET 2017) and the 2018 EAI International Conference on Robotic Sensor Networks (ROSENET 2018). The theme of ROSENET 2017 and ROSENET 2018 was "Cognitive Internet of Things for Smart Society." This proceedings highlights selected papers presented at the 1st/2nd EAI International Conference on Robotic Sensor Networks, held in Kitakyushu, Japan. Today, the integration of artificial intelligence and internet of things has become a topic of growing interest for both researchers and developers from academic fields and industries worldwide, and artificial intelligence is poised to become the main approach pursued in next-generation IoTs research.

The rapidly growing number of artificial intelligence algorithms and big data devices has significantly extended the number of potential applications for IoT technologies. However, it also poses new challenges for the artificial intelligence community. The aim of this conference is to provide a platform for young researchers to share the latest scientific achievements in this field, which are discussed in these proceedings.

The technical program of ROSENET 2017 and ROSENET 2018 consisted of 19 full papers from 39 submissions, including 18 papers in main track and 1 invited paper in special session "Artificial Tactile Sensing and Haptic Perception." Aside from the high-quality technical paper presentations, the technical program also featured two keynote speeches. The five keynote speakers were Prof. Seiichi Serikawa, Prof. Hyoungseop Kim, Prof. JooKooi Tan from Kyushu Institute of Technology, Japan, Prof. Yujie Li from Fukuoka University, Japan, and Prof. Min Chen from Huazhong University of Science and Technology, China.

Coordination with the steering chair, Imrich Chlamtac, was essential for the success of the conference. We sincerely appreciate his constant support and guidance. It was also a great pleasure to work with such an excellent organizing committee team for their hard work in organizing and supporting the conference: in particular, the Technical Program Committee, led by our TPC Co-chairs, Dr. Shenglin Mu, Dr. Jože Guna, and Dr. Shota Nakashima who have completed the peer-review process of technical papers and made a high-quality technical program. We are also grateful to Conference Manager, Dominika Belisova, for her support and all the authors who submitted their papers to the ROSENET 2017 and ROSENET 2018 conferences and special sessions.

We strongly believe that ROSENET conferences provide a good forum for all researchers, developers, and practitioners to discuss all science and technology aspects that are relevant to Robotics and Cognitive Internet of Things. We also expect that the future ROSENET conferences will be as successful and stimulating as indicated by the contributions presented in this volume.

Kitakyushu, Japan   Huimin Lu
Fukuoka, Japan   Li Yujie

Conference Organization

Steering Committee	
Imrich Chlamtac	European Alliance for Innovation, Gent, Belgium
Huimin Lu	Kyushu Institute of Technology, Japan
Organizing Committee	
General Chair	
Hyoungseop Kim	Kyushu Institute of Technology, Japan
Shota Nakashima	Yamaguchi University, Japan
Huimin Lu	Kyushu Institute of Technology, Japan
Technical Program Chair	
Yin Zhang	Zhongnan University of Economics & Law, China
Dong Wang	Dalian University of Technology, China
Shenglin Mu	Ehime University, Japan
Joze Guna	University of Ljubljana, Slovenia
Yujie Li	Yangzhou University, China
Publication Chair	
Kuan-Ching Li	Providence University, Taiwan
Local Chairs	
Tomoki Uemura	Kyushu Institute of Technology, Japan
Shingo Aramaki	Yamaguchi University, Japan
Workshop Chair	
Guangxu Li	Tianjin Polytechnic University, China
Exhibits Chair	
Kauhiro Hatano	Kyushu Institute of Technology, Japan
Demos Chair	
Xing Xu	University of Electronic Science and Technology of China, China

Posters Chair	
Tongwei Ren	Nanjing University, China
Publicity Chairs	
Li He	Qualcomm Inc., USA
Shenglin Mu	Ehime University, Japan
Quan Zhou	Nanjing University of Posts and Telecomm., China
Zongyuan Ge	IBM Inc., Australia
Conference Manager	
Alzbeta Mackova	EAI—European Alliance for Innovation
Technical Program Committee	
Rushi Lan	University of Macau, Macau (Chair)
Hu Zhu	Nanjing University of Posts and Telecommunications, China (Co-chair)
Jihua Zhu	Xi'an Jiaotong University, China
Xin Jin	Beijing Electronic Science and Technology Institute, China
Baoru Han	Hainan Software Profession Institute, China
Mei Wang	Xi'an University of Science and Technology, China
Mingwei Cao	Hefei University of Technology, China
Narisha	Harbin University of Science and Technology, China
Min Jiang	Jiangnan University, China
Baoru Han	Hainan Software Profession Institute, China
Peng Geng	Shijiazhuang Tiedao University, China
Shuaiqi Liu	Hebei University, China
Lei Mei	California Research Center, Agilent-Santa Clara, USA

Contents

New Tuning Formulas: Genetic Algorithm Used in Air Conditioning Process with PID Controller

Xiaoli Qin, Hao Li, Weining An, Hang Wu, and Weihua Su

1 Introduction

Proportional–integral–derivative (PID) controllers are the most popular controllers used in industry [1–3], and [4] have shown that 90% of control loops are of PI or PID structure. The air conditioning process is to adjust the various air parameters and these controllers are of the PID structure generally. Several parameters of the PID controller can be adjusted and the control loop can perform better if the parameters are chosen properly. The procedure of finding the controller parameters is called tuning [5]. Tuning methods research has made a rapid development in the past decades, such as Ziegler–Nichols method [6, 7], Haalman and tuning method [8, 9], Cohen–Coon (C-C) method [10, 11], and other methods [12, 13].

In this study, the reverse scheme is used in the derivation of new tuning formulas, seeking directly the relation between optimization results and model parameters by dimensionless processing. The new tuning formulas proposed are of simpler structure and similar control performance compared with complex optimization algorithms. The performance of the new tuning formulas is tested for different representative air conditioning processes compared with the Z-N method, C-C method, and AMIGO method in terms of the set-point response, the load disturbances response, and robustness. Performance analysis reveals that the new tuning formulas can provide consistently improved performance for air conditioning processes compared to other tuning methods.

X. Qin · H. Li · W. An · H. Wu · W. Su (✉)
Institute of Medical Equipment, Tianjin, China

© Springer Nature Switzerland AG 2020
H. Lu, L. Yujie (eds.), *2nd EAI International Conference on Robotic Sensor Networks*, EAI/Springer Innovations in Communication and Computing,
https://doi.org/10.1007/978-3-030-17763-8_1

2   PID Controller Structure Design and Parameters Tuning

A qualified controller should perform well in set-point response, load disturbance response, and robustness. However, those different performances are coupled. In this study, we use reverse scheme. Apply complicated optimization algorithms to a batch including various representative air conditioning processes. Try to find the relation between the normalized controller parameters and plant parameters, and present the new tuning formulas that can be of simple and convenient and with similar control performance compared to complex optimization algorithms.

2.1   Two Degrees-of-Freedom PID Control Structure

Good rejection of load disturbance will increase set-point response deviation, so we use the two degrees-of freedom structure to make the problem be decoupled. A simple way to achieve this two degrees-of-freedom structure is to use set-point weight as shown in Fig. 1.

The controller is described by

$$u(t) = \gamma K \left[e_p + (1/T_i) \int e \, dt + T_d \, de_d/dt \right] \tag{1}$$

where

$$e_p = b y_{sp} - y, e = y_{sp} - y, e_d = c y_{sp} - y \tag{2}$$

where y is the output of the process, y_{sp} is the set-point, and b and c are the set-point weighting parameters, When the process gain is greater or less than zero, then $\gamma = 1$ or $\gamma = -1$.

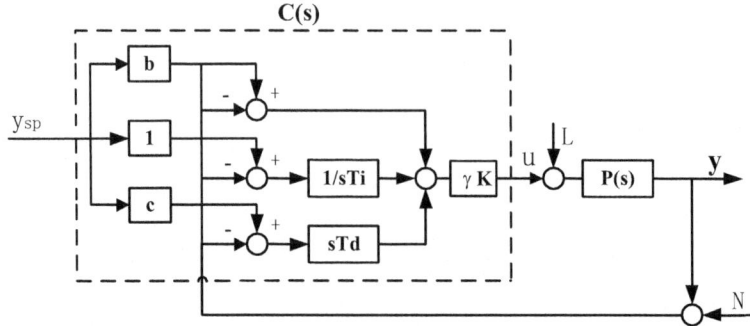

Fig. 1 Set-point weighted PID controller

2.2 Solution of Optimization Problem Using Genetic Algorithm

The optimization problem can be stated as follows:

Find controller parameters that minimize *IAE* to the constraints that the closed-loop system is stable and that the robustness index $M_s < m$. In this study, $m = 2$, the standard value of M_s [5]. Genetic algorithm is a stochastic optimization algorithm that was originally motivated by the mechanisms of natural selection and evolutionary genetics, and has found extensive application in solving global optimization problem.

The optimization procedure is summarized as follows:

1. Encoding and initialization of population

    Real-number encoding [14] was used in this study, the PID parameters K, T_i, T_d that we need to tune compose a chromosome (K, T_i, T_d), where $0 < K < K_u$, $0 < T_i < T_u$, and $0 < T_d < T_u$, where K is the proportional gain, T_i is the integral time, and T_d is the derivative time. K_u is the ultimate gain and T_u is the ultimate period of the process.

    The initial population generation is randomly chosen within these ranges.
2. Fitness and cost function

    In this study, the cost function and fitness function $J(K, T_i, T_d)$ were defined as

$$J(K, T_i, T_d) = \text{IAE}_{\text{load}} \tag{3}$$

$$F(K, T_i, T_d) = \frac{1}{J(K, T_i, T_d)} \tag{4}$$

Our objective was to search (K, T_i, T_d), in which $J(K, T_i, T_d)$ is minimized to the robustness constraints. A better PID controller will be of better fitness. The GA generates better offspring to improve the fitness, approaching the optimal PID parameters.

If any of following conditions is true, $F(K, T_i, T_d) = 0$, this chromosome will be eliminated:

(a) A system with PID parameters K, T_i, T_d is unstable.
(b) $M_s > 2$.
(c) Any PID parameters K, T_i, T_d are negative.

3. Elimination and duplication

    In a generation with $N = 200$ individuals, we calculated the fitness of every individual and defined the (subsistence power):

$$SP_i = N \frac{F_i}{\sum\limits_{i=1}^{N} F_i} \tag{5}$$

When SP < 1, those individuals are eliminated, and those individuals with higher SP will duplicate themselves to occupy extra space, keeping the population size N unchanged. Hence, the individual with high fitness is more likely to be chosen for crossover and mutation.

4. Crossover and mutation

Crossover exchanges the information of two chromosomes. First, we set a crossover probability $P_c = 0.8$. Second, for two arbitrary chromosome, the procedure gives a random number N_c, where $N_c, P_c \in [0, 1]$. If $N_c < P_c$, the crossover between the two chromosomes will happen. The crossover rule is

$$A : MA_{n+1} = \lambda MA_n + (1 - \lambda) MB_n \tag{6}$$

$$B : MB_{n+1} = \lambda MB_n + (1 - \lambda) MA_n \tag{7}$$

where λ is a random number between 0 and 1, and the MA_n and MB_n represent the PID parameters K, T_i, T_d in A and B chromosome of n generations, respectively. Third, the second step is repeated until the offspring's components are completed.

Mutation is a way to add new genetic material to the population to avoid getting stuck at a local optimal. The mutation procedure is similar to crossover, and the mutation probability is $P_m = 0.2$. When a random number $N_m < P_m$, N_m, $P_m \in [0, 1]$, the chromosome will undergo mutation. The mutation rule is

$$C : MC_n = MC_{mean} + MC_{range} (\eta - 0.5) \tag{8}$$

where η is a random number between 0 and 1, and the MC_{mean} and MC_{range} represent the mean and the range of the PID parameters K, T_i, T_d in C chromosome of n generations, respectively.

Through the crossover and mutation, the new generation is produced, then compute the fitness of every chromosome and repeat elimination and reproduction until the final generation of $n = 100$ is attained. The PID parameters in the chromosome of largest fitness in the final generation are the result of optimization.

The optimization procedure is summarized as follows:

Step 1: Compute the ultimate gain and ultimate period of the air conditioning process.

Step 2: Generate the initial population of size N.

Step 3: Calculate the fitness value for each chromosome composed by the PID parameters.

Step 4: Through elimination, duplication, crossover, and mutation, produce a new population.

Step 5: If the final generation is attained, stop the search, else go to *Step* 3.
Step 6: Find the chromosome of largest fitness in the final generation and output its
PID parameters as the optimization result.

3   Derivation of the New Tuning Formulas

3.1   Test Batch

The process information is available for air conditioning process in terms of a first-order plus dead time (FOPDT) model as given in (1). The model parameters are normalized by an important parameter called the relative dead time $\tau = L/(L + T)$, where L/T is also an important index in controller parameters tuning. Cohen and Coon called L/T the self-regulating index [5]. It is also called the controllability index [8]; Astrom and Hagglund use the relative dead time τ instead of L/T for the advantage that the parameter is bounded to the region [0, 1] [13].

The test batch includes 60 different FOPDT models representing the various types of air conditioning process. K_p is assumed to be the standard 1. T and L are the values taken from the sets $T_{\text{set}} = (600, 700, \ldots, 1500, 1600)$ and $T_{\text{set}} = (100, 200, \ldots, 800, 900)$ and $\tau \approx (0\text{--}0.6)$ correspondingly.

3.2   Apply Optimization Procedures

In this study, the population size N and the maximum allowable generation number n were chosen as 200 and 100, respectively. Search the best PID controller parameters for every model in the batch using the optimization procedure described in Sect. 2.

3.3   New Tuning Formulas

The PID controller parameters were normalized to six types of dimension free parameters, i.e., KK_p, aK, T_i/T, T_i/L, T_d/T, and T_d/L. The least-squares method is used to find the best function match, completing the curve fitting. Three PID parameters used the same form of expression, i.e., exponential function. The new formulas are summarized up in (9):

$$F(\tau) = ae^{b\tau} + ce^{d\tau} \tag{9}$$

The corresponding function coefficients of the new formulas (9) are given in Table 1.

Table 1 Coefficients of the new tuning formulas	$F(\tau)$	a	b	c	d
	KK_p	11.09	−5.598	0.282	1.44
	T_i/L	52.01	−35.54	2.844	−1.799
	T_d/T	0.105	2.741	−0.109	−0.584

Table 2 The PID parameters of different tuning methods	Methods	K	T_i	T_d	b	c
	Z-N	4.83	580	145	1	1
	C-C	5.63	674.27	103.55	1	1
	AMIGO	3.59	491.16	122.05	0	0
	New-rule	3.54	699.63	97.964	0.17	0.625

The new formulas can give similar control performance compared to complex optimization algorithms while be of simple form.

4 Results

Effectiveness of the proposed new tuning formulas is verified by simulation experiments on the temperature conditioning process. Relative dead time τ of the three processes is in different region representing different air conditioning processes. Performance of the proposed new tuning formulas is compared with those of Ziegler–Nichols (Z-N) method, Cohen–Coon (C-C) method, and AMIGO method with respect to set-point response, load disturbance response, and robustness using the indexes: $M_p(s)$ and t_s, IAE and $M_p(L)$, M_s, respectively.

For our simulation experiments, we consider a cooling conditioning process. Transfer function of this process is given by

$$G(s) = -\frac{1.3e^{-290s}}{1520s + 1} \tag{10}$$

The relative dead time of this process is $\tau = 0.16$. The controller parameters provided by different methods are shown in Table 2. Due to the minus process gain, $\gamma = -1$.

To better demonstrate the performance differences between other different methods in the figure, Z-N method was not shown due to its unacceptable controller output and bad time responses. The time responses and Nyquist curves of the loop transfer function designed by different tuning method are shown in Figs. 2 and 3, respectively. Figure 2 exhibits the responses of the system to changes in set-point and load disturbances. Load disturbance attenuation of the three methods is similar, but the considerably improved performance of the new tuning formulas over C-C, even better than AMIGO can be seen during the set-point response. The variation of control actions for different tuning methods corresponding to response are also shown in Fig. 2. The control signal of new tuning formulas is the smoothest and

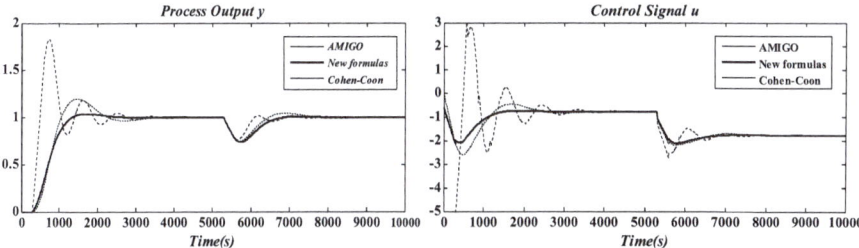

Fig. 2 Responses to a unit step change at time 0 and a load step at time 5000 for PID controller designed by different tuning methods

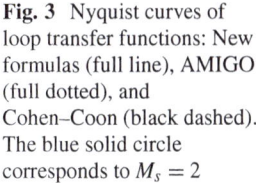

Fig. 3 Nyquist curves of loop transfer functions: New formulas (full line), AMIGO (full dotted), and Cohen–Coon (black dashed). The blue solid circle corresponds to $M_s = 2$

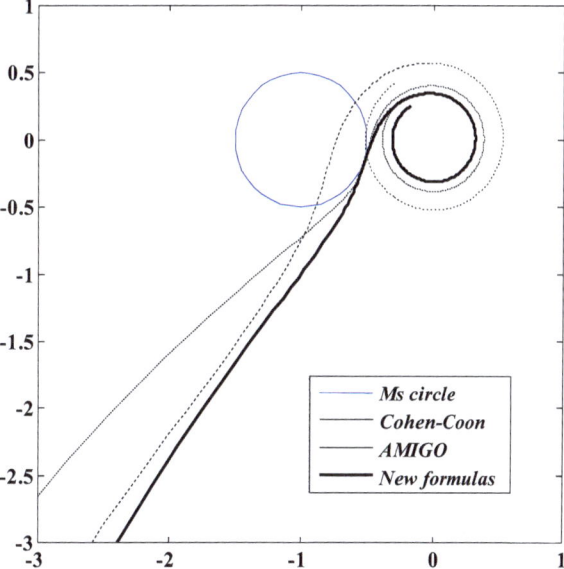

minimum change. The Nyquist curve of a qualified system should be guaranteed to stay out of the M_s circle. Figure 3 exhibits the unqualified robustness in the system designed by C-C method.

To compare the performance between different tuning methods more clearly, quantitative analysis is provided in Table 3. The new tuning formulas perform qualified in the different aspects, doing a good trade-off between different performances. $M_p(s)$ of Z-N and C-C methods seems to be very large (about 61% and 82.5%), which may not be acceptable in many situations. The new tuning formulas reduce the $M_p(s)$ by more than 20 times compared to C-C tuning method. Observe that AMIGO method provides four times overshoot with no improvement in the load regulation with respect to new tuning formulas. The IAE index for Z-N and C-C tuning methods seems to be very good (about 130) compared to new formulas (about

Table 3 The PID parameters of different tuning methods

Methods	Set-point		Load disturbance		Robust
	$M_p(s)$ (%)	$t_s = (5\%)$ (s)	LAE	$M_p(L)$	M_s
Z-N	61	1679	139.16	0.232	3.4
C-C	82.5	2200	132.03	0.236	4
AMIGO	20	2070	202.76	0.254	2
New-rule	3.9	1185	203.46	0.263	2

200), but at the expense of the robustness. The robustness index is about double the allowable value $M_s = 2$. The system is fragile and does not have a real value.

5 Conclusion

A reverse scheme is used in the derivation of the new tuning formulas, seeking the relation between the optimization results and model parameters directly by dimensionless processing. The new tuning formulas are of simple structure and similar control performance compared to complex optimization algorithms. Effectiveness and usefulness of the proposed tuning formulas has been tested on different representative air conditioning processes. The new tuning formulas have shown consistently improved performance both in set-point and load disturbances response compared to typical tuning methods for air conditioning processes.

Acknowledgements This work is financially supported by the AMMS Youth Innovation Foundation No. 2017CXJJ09.

References

1. Koivo, H. N., & Tanttu, J. T. (1991). Tuning of PID controllers: Survey of Siso and Mimo techniques. In *Intelligent tuning and adaptive control* (pp. 75–80). Singapore: IFAC Intelligent Tuning Adaptive Control Symposium.
2. Ingimundarson, A., & Hägglund, T. (2002). Performance comparison between PID and dead-time compensating controllers. *Journal of Process Control, 12*(8), 887–895.
3. Åström, K. J., & Hägglund, T. (2001). The future of PID control. *Control Engineering Practice, 9*(11), 1163–1175.
4. Yamamoto, S. (1991). Present status and future needs: The view from Japanese industry. In *Proceedings of 4th International Conference on Chemical Process Control-CPC IV, CACHE-AIChE* (pp. 1–28).
5. Åström, K. J., & Hägglund, T. (1995). *PID controllers: Theory, design and tuning*. Research Triangle Park: Instrument Society of America.
6. Ziegler, J. G. (1942). Optimum settings for automatic controllers. *Transactions of the ASME, 64*(2B), 759–768.

7. Åström, K. J., & Hägglund, T. (2004). Revisiting the Ziegler–Nichols step response method for PID control. *Journal of Process Control, 14*(6), 635–650.
8. Åström, K. J., & Hägglund, T. (2006). *Advanced PID control*. Research Triangle Park: ISA.
9. Hägglund, T., & Åström, K. J. (2010). Revisiting the Ziegler-Nichols tuning rules for pi control. *Asian Journal of Control, 6*(4), 469–482.
10. Chilali, M., & Gahinet, P. (1996). H∞ design with pole placement constraints: An LMI approach. *IEEE Transactions on Automatic Control, 41*(3), 358–367.
11. Shih, Y., & Chen, C. (1974). On the weighting factors of the quadratic criterion in optimal control. *International Journal of Control, 19*(5), 947–955.
12. Skogestad, S. (2003). Simple analytic rules for model reduction and PID controller tuning. *Journal of Process Control, 13*(4), 291–309.
13. Panagopoulos, H. (1998). *Design of PI controllers based on non-convex optimization*. Oxford: Pergamon Press.
14. Gen, M., & Cheng, R. (2000). *Genetic algorithms and engineering optimization* (Vol. 7). New York: Wiley.

A Multi-Level Thresholding Image Segmentation Based on an Improved Artificial Bee Colony Algorithm

Xingyu Xia, Hao Gao, Haidong Hu, Rushi Lan, and Chi-Man Pun

1 Introduction

Image segmentation technique is an attempt to find interesting parts in an image, which is an important component in image processing, video processing, and analysis [1–3]. As a traditional image segmentation algorithm, thresholding methods segment a digital image into multiple parts. They have been divided into two categories: bi-level segmentation and multi-level segmentation. The former category means an image is divided into two subdivisions, which uses one gray value to represent its threshold. The multi-level segmentation method discriminates several distinct subdivisions from a digital image which uses more than two thresholds.

X. Xia
The Institute of Advanced Technology, Nanjing University of Posts and Telecommunications, Nanjing, China

H. Gao (✉)
The Institute of Advanced Technology, Nanjing University of Posts and Telecommunications, Nanjing, China

Department of Computer and Information Science, University of Macau, Macau SAR, China

H. Hu
Beijing Institute of Control Engineering, Beijing, China

R. Lan
Key Laboratory of Intelligent Processing of Computer Image and Graphics, Guilin University of Electronic Technology, Guilin, China

C.-M. Pun
Department of Computer and Information Science, University of Macau, Macau SAR, China

© Springer Nature Switzerland AG 2020
H. Lu, L. Yujie (eds.), *2nd EAI International Conference on Robotic Sensor Networks*, EAI/Springer Innovations in Communication and Computing, https://doi.org/10.1007/978-3-030-17763-8_2

11

As a representative threshold-based segmentation method, Otsu [4] has attracted many people to do further research. Based on previous findings, the Otsu method can be treated as a maximum optimization problem. But a traditional exhausted searching method expends too much computational time to endure especially on multi-level threshold problems. As a population-based algorithm, evolutionary algorithms (EAs) search a potential solution space by using multiple individuals, which means they show fast computational ability than the traditional exhausted searching methods.

Artificial bee colony (ABC) [5] has been demonstrated as an efficient EA, due to its strong global search ability and steady robustness. Compared with the other EAs, ABC has two specific characteristics. The first one is a one-dimensional search strategy which means bees in ABCs search the potential solutions in one by one dimension. Due to this property, the ABCs demonstrate good performances especially on separable functions, in which the change of one variable could not cause changes of other variables. Another characteristic of ABC is its scout bee search strategy. It reinitializes the individual which has not improved in a defined iteration. Compared with other EAs, this strategy guarantees the global search ability of ABC. In this paper, we first investigate the property of Otsu function. Then we design an improved ABC algorithm specific for the Otsu segmentation function.

The following sections are organized as the following sequence: In Sect. 2, a detailed information of the traditional artificial bee colony algorithm is presented. Otsu for multi-level thresholding image segmentation problem is elaborately presented in Sect. 3. Based on analyzing the characteristics of the Otsu, we propose a new ABC algorithm specially designed for this thresholding segmentation algorithm in Sect. 4. The experiments on Berkeley image database are conducted in Sect. 5 and an integrate conclusion is presented in Sect. 6.

2 ABC Algorithm

Artificial bee colony algorithm was first proposed by Karaboga in 2005 [5], and its essential principle is to simulate the bee's foraging behavior, with large amount of nectar to find food sources. The algorithm consists of three types of bees, which are employed bees, onlooker bees, and scout bees, each of which corresponds to a different search task. Each phase can be described as follows:

Initialization At first, all the bees were randomly initialized in a potential solution space. The maximum boundary and minimum boundary of the search space are represented by upper and lower, respectively. The number of food source is NP, then the D-dimensional vector as $X_i = (x_{i1}, x_{i2}, \cdots, x_{iD})$ indicates the ith food source position and x_{ij} denotes the jth variable of X_i, $i = 1, 2, \ldots, NP, j = 1, 2, \ldots, D$. The updating equation of initialization can be expressed as

$$x_{ij} = \text{Lower}_{ij} + \left(\text{Upper}_{ij} - \text{Lower}_{ij}\right) * \text{rand}(0, 1) \tag{1}$$

where rand(0, 1) is a real number selected within (0, 1) randomly and uniformly.

Employed Bees The employed bees will exploit the potential food sources by combining the previous experience and information of a randomly selected neighbor bee with Eq. (2)

$$v_{ij} = x_{ij} + \phi\left(x_{ij} - x_{kj}\right) \tag{2}$$

where V_i is the candidate solution of i and v_{ij} represents its jth variable; k represents the number of a neighbor bee in $[1, \text{NP}]$; and ϕ is a uniformly distributed real value randomly selected in interval of $[-1, 1]$.

Suppose that it is a minimum optimization problem, the fitness value F_i is calculated using Eq. (3)

$$F_i = \begin{cases} 1/(1 + f_i) & \text{if } f_i \geq 0 \\ 1 + |f_i| & \text{if } f_i < 0 \end{cases} \tag{3}$$

where f_i is the optimization function value.

If the fitness of candidate solution V_i gets better than the current solution, the current food source X_i will be updated by the candidate solution V_i; otherwise, X_i will remain unchanged.

Onlooker Bees The selection probability of each employed bee to use in the onlooker bees is calculated

$$p_i = \frac{F_i}{\sum_1^{\text{NP}} F_i} \tag{4}$$

According to the probability calculated by Eq. (4), onlooker bees use roulette strategy to select a source for intensive search with Eq. (2). The greedy selection strategy is the same as the employed bees' method.

Scout Bees When a food source stops updating for a successive iteration, it should be selected as a scout bee and reinitialized in the solution space to replace the previous value of this food source.

3 Otsu Segmentation Algorithm

Suppose there is a minimum gray level 0 and a maximum gray level L in the image which has N pixels. In the ith gray level, the number of pixels is $h(i)$ and the probability is denoted as PR_i.

$$\text{PR}_i = h(i)/N \quad \text{and} \quad N = \sum_{i=0}^{L} h(i) \tag{5}$$

The Otsu segmentation method discriminate $M - 1$ thresholds, $\{\text{th}_1, \text{th}_2, \ldots, \text{th}_{M-1}\}$ for consumers need M parts of an image. Let $C_j(j = 1, 2, \ldots, M)$ represent the ith part of gray level. Then we have C_1 for $[0, \cdots, \text{th}_1]$, $\ldots$, and C_M for $[\text{th}_M + 1, \cdots, L]$. The Otsu method determines the optimal thresholds by using the criterion of minimal intra-class deviation and maximal between-classes deviation.

The optimal thresholds $\{\text{th}_1^*, \text{th}_2^*, \ldots, \text{th}_{M-1}^*\}$ can be calculated as

$$\{\text{th}_1^*, \text{th}_2^*, \ldots, \text{th}_{M-1}^*\} = \arg\max\{\sigma_B^2 (\text{th}_1, \text{th}_2, \ldots, \text{th}_{M-1})\};$$
$$0 \le \text{th}_1 \le \text{th}_2 \le \cdots \le \text{th}_{M-1} \le L \tag{6}$$

where

$$\sigma_B^2 = \sum_{k=1}^{M} w_k{}^* \left(u_k - u_{t_k^*}\right)^2 \quad w_k = \sum_{i \in C_k} \text{PR}_i \quad u_k = \sum_{i \in C_k} i\text{PR}_i/w_k, k = 1, \ldots, M \tag{7}$$

Therefore, Otsu method can be regarded as the maximum problem. Due to the Otsu method of traditional computing strategy is particularly time-consuming, EA based Otsu segmentation method should accelerate its rate of finding the optimal thresholds.

4 An Improved ABC Algorithm

4.1 Analysis on the Otsu

A good optimization algorithm is that specifically designs for a defined optimized problem. As presented in the above, the Otsu algorithm is designed for finding thresholds $\{\text{th}_1^*, \text{th}_2^*, \ldots, \text{th}_{M-1}^*\}$ to maximize the σ_B^2. In this function, the two components w_k and u_k are constant for a specific image. Then σ_B^2 could further be described as follows:

$$\sigma_B^2 = \sum_{k=1}^{M} A^*(B - X_k)^2 \tag{8}$$

In this paper, we use the differential grouping (DG) [6] to investigate the linkage between the variables in Otsu method. It is a variable grouping algorithm that can detect the interacting variables in a black-box optimization problem. DG is based on the following theorem:

Theorem 1 Let $f\left(\overrightarrow{x}\right)$ be an additively separable function. $\forall a, b_1 \neq b_2, \delta \in R^1$, $\delta \neq 0$, if the following condition holds:

$$\Delta_{\delta, x_p}[f]\left(\overrightarrow{x}\right)\big|_{x_p = a, x_q = b_1} \neq \Delta_{\delta, x_p}[f]\left(\overrightarrow{x}\right)\big|_{x_p = a, x_q = b_2} \tag{9}$$

then x_p and x_q interact, where

$$\Delta_{\delta, x_p}[f]\left(\overrightarrow{x}\right) = f\left(\ldots, x_p + \delta, \ldots\right) - f\left(\ldots, x_p, \ldots\right) \tag{10}$$

refers to the forward difference of f with respect to variable x_p with interval δ.

Theorem 1 means that two variables x_p and x_q are separable if Eq. (10) gets the same results when it is evaluated with any two different values of x_q.

According to this theory and experimental results, we could find that these variables of Otsu function are separable. Since one of the main properties of ABC is its one-dimensional search strategy, we propose an improved ABC algorithm based Otsu image segmentation algorithm.

4.2   An Improved ABC

It has bee proven that ABC shows power global but poorer local search abilities, which means it can get closer to a potential solution but can't get precise results. Furthermore, we can easily find that the Otsu function is a unimodal function, which means it emphasis on the local search ability of an optimization algorithm. Combining the property of the Otsu function and the ABC, we propose a new ABC algorithm with an improved scout bee strategy. In this algorithm, for accelerating the convergence rate of bees, we first introduce the experience of population into onlooker bees. Then different from the traditional copy, the scout bee strategy is triggered for making local searches when it has not been improved for a limitation iteration.

The new updating equation of onlooker bee is listed as follows:

$$v_{ij} = x_{ij} + c1^*\left(x_{ij} - x_{kj}\right) + c2^*\left(\text{gbest} - x_{ij}\right) \tag{11}$$

where gbest represents the experience of population. $c1$ and $c2$ represent the two Gaussian operators whose mean value and standard deviation are set as 0 and 0.2, respectively.

Since the role of scout bees is used to guarantee the global search ability of population in the traditional ABC, it should play an important role in a multimodal function. But for a simple function, the scout bee strategy could only guarantee the reliability of a found solution but the precise of the solution maybe should be weakened and the function evaluation in the original copy is wasted. For having more chance to making further precise search, the new updating equation of scout bee is described as follows:

$$x_i = x_i \pm 5 \tag{12}$$

where x_i represents the scout bee position.

We should note that the value of a potential solution is integer for a thresholding image segmentation problem. Then all the updating equations of our proposed algorithm use cell operation in MATLAB to get appropriate solution.

5 Experiment

5.1 Experimental Setting

For evaluating the performance, we compare six popular image segmentation algorithms based on EAs with our algorithm. These algorithms are PSO [7], QPSO [8], DE [9], ABC [10], MABC [11], and I_ABC [12] which are specifically designed for an image segmentation problem. The population size is set as 40. The maximum iteration is set as 60 on $M - 1 = 3$ and 100 on $M - 1 = 5, M - 1 = 6$, respectively. The particular parameters for the compared algorithm are set as their own papers. The tested images are adopted from the Berkeley image database (http://www.eecs.berkley.edu/Research/Projects/CS/vision/bsds/). For the limited length of paper, we only demonstrate the result of the former five images.

5.2 Experimental Results

Table 1 list the results of the compared algorithms, the first and second row lists their mean values and standard deviations separately. Figure 1 shows the five-dimensional segmentation results. From the results, we could find that our algorithm gets more favorable results on most thresholding images. When $M - 1 = 3$, we find that the QPSO shows more stable results than the other algorithms. It is mainly due to its gbest component and quantum-behaved theory. Our algorithm also gets the best

Table 1 The compared results of different EA based Otsu image segmentation algorithm

	$M-1$	PSO	QPSO	DE	ABC	MABC	IABC	SABC
1	3	8.126e3	**8.128e3**	**8.128e3**	8.126e3	**8.128e3**	8.123e3	**8.128e3**
		5.107	2.73e−12	0.5803	6.639	**2.728e−12**	9.955	0.939
	5	9.345e3	9.386e3	9.368e3	9.386e3	9.386e3	8.903e3	**9.389e3**
		40.694	14.627	19.182	12.711	6.2818	2042.7	**6.03**
	6	9.6e3	9.671e3	9.664e3	9.685e3	9.228e3	8.7e3	**9.709e3**
		61.32	35.48	23.24	22.46	2117.1	2899	**19.68**
2	3	8.408e3	**8.415e3**	8.411e3	**8.415e3**	**8.415e3**	8.409e3	**8.415e3**
		8.761	**0.2055**	4.326	1.532	0.3367	6.612	1.178
	5	9.931e3	9.965e3	9.935e3	9.969e3	9.472e3	9.948e3	**9.974e3**
		31.726	13.248	20.797	4.7126	2173	13.47	**0.731**
	6	1.029e4	1.037e4	1.032e4	1.037e4	1.037e4	1.034e4	**1.038e4**
		55.36	12.27	28.202	7.668	16.724	26.23	**6.697**
3	3	6.098e3	**6.1e3**	6.097e3	6.099e3	6.099e3	6.097e3	6.099e3
		1.8336	**1.82e−12**	3.001	1.677	0.414	3.302	1.106
	5	**7.267e3**	7.188e3	7.166e3	7.183e3	6.826e3	7.178e3	7.189e3
		16.116	4.726	14.848	4.304	1566	8.818	**2.349**
	6	7.427e3	7.48e3	7.449e3	7.481e3	7.48e3	7.465e3	**7.486e3**
		38.501	9.592	12.747	6.569	5.775	13.793	**3.56**
4	2	5.216e3	**5.217e3**	5.215e3	**5.217e3**	**5.217e3**	5.215e3	**5.217e3**
		1.572	**0.0032**	1.324	0.565	0.0068	2.497	0.99
	5	5.947e3	5.957e3	5.946e3	5.956e3	5.956e3	5.66e3	**5.962e3**
		9.0655	5.421	8.934	3.7508	1298.3	5.672	**1.214**
	6	6.146e3	6.158e3	6.136e3	6.156e3	5.85e3	6.15e3	**6.162e3**
		12.128	5.37	8.981	3.3629	1342	5.545	**2.375**
5	3	**9.899e3**	**9.899e3**	9.896e3	**9.899e3**	**9.899e3**	9.401e3	**9.899e3**
		0.8144	**0.3388**	1.8229	0.901	0.574	2156.7	0.8155
	5	1.142e4	**1.143e4**	1.14e4	**1.143e4**	**1.143e4**	1.141e4	**1.143e4**
		12.714	5.372	13.71	4.907	5.473	8.91	**1.652**
	6	1.185e4	1.187e4	1.184e4	1.188e4	1.190e4	1.185e4	**1.191e4**
		36.819	26.992	29.218	18.568	22.021	19.815	**16.642**

mean values on $M-1=3$ and shows the best results especially when $M-1=5$, $M-1=6$. Compared with the other ABC variants, the gbest component and the Gaussian operator enable the onlooker bees to focus on a potential valuable region. Furthermore, the new scout bee strategy also improves the local search when the stagnation happens.

Fig. 1 Five-dimensional segmentation results

6 Conclusion

Based on analyzing the property of the Otsu function, this paper put forward an ABC algorithm with an improved scout bee strategy to make precise searches. Experimentations on benchmark images have demonstrated its efficiency and effectiveness. Our prospective research will focus on further enhancing the performance of an image segmentation method based on EAS.

Acknowledgements The authors acknowledge the support from National Nature Science Foundation of China (No. 61571236, 61533010, 61320106008, 61602255) and Postgraduate Research & Practice Innovation Program of Jiangsu Province (KYCX17_0795).

References

1. Mylonas, S. K., Stavrakoudis, D. G., Theocharis, J. B., et al. (2016). A local search-based GeneSIS algorithm for the segmentation and classification of remote-sensing images. *IEEE Journal of Selected Topics in Applied Earth Observations and Remote Sensing, 9*(4), 1470–1492.
2. Pinheiro, M., & Alves, J. L. (2015). A new level-set-based protocol for accurate bone segmentation from CT imaging. *IEEE Access, 3*, 1894–1906.
3. Dang, C., Gao, J., Wang, Z., et al. (2015). Multi-step radiographic image enhancement conforming to weld defect segmentation. *IET Image Processing, 9*(11), 943–950.

4. Otsu, N. (1979). A threshold selection method from gray-level histograms. *IEEE Transactions on Systems, Man, and Cybernetics, 9*(1), 62–66.
5. Karaboga, D. (2005). *An idea based on honey bee swarm for numerical optimization. Technical report-tr06*. Kayseri: Erciyes University.
6. Omidvar, M. N., Li, X., Mei, Y., & Yao, X. (2014). Cooperative co-evolution with differential grouping for large scale optimization. *IEEE Transactions on Evolutionary Computation, 18*(3), 378–393.
7. Akay, B. (2013). A study on particle swarm optimization and artificial bee colony algorithms for multilevel thresholding. *Applied Soft Computing, 13*(6), 3066–3091.
8. Cao, L. L., Ding, S., Fu, X. W., et al. (2016). Otsu multilevel thresholding segmentation based on quantum particle swarm optimisation algorithm. *International Journal of Wireless and Mobile Computing, 10*(3), 272–277.
9. Sarkar, S., & Das, S. (2013). Multilevel image thresholding based on 2D histogram and maximum Tsallis entropy—A differential evolution approach. *IEEE Transactions on Image Processing, 22*(12), 4788–4797.
10. Kumar, S., Kumar, P., Sharma, T. K., et al. (2013). Bi-level thresholding using PSO, artificial bee colony and MRLDE embedded with Otsu method. *Memetic Computing, 5*(4), 323–334.
11. Horng, M. H. (2011). Multilevel thresholding selection based on the artificial bee colony algorithm for image segmentation. *Expert Systems with Applications, 38*(11), 13785–13791.
12. Bhandari, A. K., Kumar, A., & Singh, G. K. (2015). Modified artificial bee colony based computationally efficient multilevel thresholding for satellite image segmentation using Kapur's, Otsu and Tsallis functions. *Expert Systems with Applications, 42*(3), 1573–1601.

Dynamic Consolidation Based on *K*th-Order Markov Model for Virtual Machines

Na Jiang

1 Introduction

With the rapid development of the Internet, a large amount of scientific data [1, 2] and business data on the computing power of the demand are far greater than its own deployment of the data center's computing power. It is because of this demand that the rapid development of cloud computing technology is promoted. Cloud computing is a dynamically scalable computing method of computer resources virtualization, as it provides services to users [3], especially for computation-sensitive scenarios [4–6]. This mode elastically supports computer resource allocation on demand, having a generally dynamic expansion and distributed characteristics. The cloud computing environment dependent on these cloud data centers has also led to great power consumption and CO_2 emissions. It is estimated that from 2005 to 2010, the worldwide power resources in data center consumption rose by 56%, accounting for 1.1–1.5% of global electricity consumption in 2010. Unless the current routine resource management program is changed to achieve efficient use of energy, data center energy consumption will continue to grow rapidly [7].

With the current problems facing cloud data centers, CPUs and other resources typically use only 10–50% of total resources, but this configuration reserved manner led to a massive waste of additional power resources [8–12]. In the cloud data center operation process, the virtual machine is a computing resource cloud computing platform for distribution and scheduling, and provides users with computing resources, with its own need to run on the physical host. According to the Open Compute Project Facebook page, power usage effectiveness (PUE) in the Prineville data center (OR, USA), in the fourth quarter of 2015 reached 1.09,

N. Jiang (✉)
Zhaotong University, Zhaotong, China
e-mail: 27805044@qq.com

© Springer Nature Switzerland AG 2020
H. Lu, L. Yujie (eds.), *2nd EAI International Conference on Robotic Sensor Networks*, EAI/Springer Innovations in Communication and Computing, https://doi.org/10.1007/978-3-030-17763-8_3

whereas in Forest City the PUE reached 1.08.[1] This means that computing power resource consumption accounted for about 91% of all cloud data center resource consumption. Therefore, to solve the problem of power consumption, the focus is on improving the cloud data center nodes in the calculation of the utilization of resources.

The main work of this paper focuses on the integration of virtual machine systems into OpenStack, for example, virtual machine integration mainly involves control node scheduling and computing nodes on the virtual machine in distribution and migration work, including computing nodes need to decision virtual machine. The need for migration and the migration time to avoid excessive "virtual" migration is caused by the "jitter" phenomenon and results in degradation of the overall performance of the cloud data center. The main purpose of virtual machine integration is to save energy consumption as much as possible under the conditions of the load balance of the cloud data center.

For this purpose, we studied the process model predictive Markov chain in time series, a new hybrid Markov model sequence K binding the Pearson correlation coefficient for predicting future host CPU utilization. The University of Melbourne, Australia released the CloudSim [13] cloud computing platform JAVA simulation package for simulation experiments based on the hybrid Markov host load predicted in the simulation of virtual machine migration and data center power consumption, such as the service level agreement (SLA) breach, which was compared with a traditional threshold-based average load detection algorithm, the local regression robust (LRR) detection algorithm [14]. The simulation results show that the host load forecasting algorithm proposed in this paper can effectively reduce the amount of virtual machine migration and cloud data center energy consumption.

The third part introduces the model and elaborates on the realization of the algorithm. The fourth part and the fifth parts introduce the design and results analysis of the experiment, and the sixth part summarizes the full text.

2 Related Work

The main purpose of dynamic consolidation, by considering the virtual machine a request for resources in real time, a method using real-time virtual machines migrated to refocus on fewer data assigned to the central node [15], and the idle switch node to the low-power host consumption state, thereby improving the use of physical resources and reducing energy consumption. There are two ways to allocate and integrate virtual machines into the cloud platform, namely, static and dynamic. In the static integration of virtual machines, resource utilization is often based on the historical average and user-defined performance indicators design and integrate the scheduling algorithm [16]. However, this approach assumes that the virtual machine

---

[1]http://opencompute.org/about/energy-efficiency/.

resource requirements are known in advance and do not take into account changes in virtual machine workloads.

Beloglazov and Buyya [15], who, after extensive research into the integration of virtual machine technology, sub-divided dynamic virtual machine integration issues into the following four questions:

1. Determining when a node enters a low load state so that all virtual machines can be migrated from that node, and the node can also be switched to a low power mode, such as sleep mode.
2. Determining when the node enters a high load state, so that the appropriate virtual machine should be selected and migration to other appropriate active nodes should be performed, to avoid server performance degradation.
3. Selecting the appropriate virtual machine for migration from the high loading host.
4. Finding the appropriate placement in the host of the other activity and migrating the virtual machine to the host.

At present, research into the decision regarding the high/low load state of the physical host can be divided into the following three methods:

1. Threshold-based heuristic methods for decision-making. In [17], Atwood et al., researched the live migration of virtual machines using a method based on a threshold value, determining a data center physical host in a high load state, then the virtual machine from its migration out to satisfy the load balanced purpose. This threshold-based load detection algorithm is relatively simple, but once the physical host of the workload changes dramatically, such as the host load state frequently goes into the high load state and low load state within a short period of time, the host may be busy migrating virtual machines, and excessive migration can cause host failures and performance degradation, which can seriously violate the SLA. In [18], Zhu et al. investigated a large number of dynamic virtual machine integration issues, using a heuristic static threshold set to 85% CPU usage. Once the node exceeded the threshold value, this determined that the node entered a high load state. In [19, 20], Gmach et al. carried out a similar study, analyzing the trajectory of a cloud data center workload, with the threshold set at 85% CPU usage. However, this method of setting the heuristic static threshold does not apply to the dynamic changes that occur in workloads or unknown host random variation.
2. Cyclical allocation of virtual machines without loading detection. In [21], Verma et al. considered dynamic virtual machine integration of simulation to be a packing problem. They considered the consumption of virtual machine migration, and proposed a heuristic approach to the data center power consumption being reduced to a minimum. However, they only periodically adjusted the location of the virtual machine without using any algorithm to determine the optimal placement of the virtual machine. If the host placed by the virtual machine goes into a high load state at the any time, the virtual machine may need to be migrated again. This leads to the frequent migration of virtual machines.

In [22], Weng et al. proposed a load balancing system for periodically reducing the consumption of virtual machines in a cluster allocated to the physical host detecting a high load and a low load, and redistributing the virtual machines. However, in large data centers, the number of both physical and virtual machines is very large. The redistribution of virtual machines inevitably leads to a lot of additional energy consumption.

3. Statistics-based approach through historical CPU utilization analysis. In [23] Bobroff et al., proposed a server load prediction algorithm based on a time series analysis of the historical data of the host, but the algorithm was too complex, and the time and space complexity too high. Huang [24], who considered the volatility and dynamic resource conflict of each virtual machine workload on the cloud computing system based on historical data of virtual machine workloads, used an auto-regressive integrated moving average (ARIMA) model to forecast future virtual machine resource requests. ARIMA is a sort of timing prediction model, which can predict the resource utilization rate of physical host nodes at the next time point using historical data. However, in the study of virtual machine integration, it is often only necessary to determine the state of the CPU, and to accurately predict its usage. Often, greater error leads to a longer running time. In [15], Beloglazov and Buyya studied the Markov chain model in predicting the future, doing a lot of the workload of the host, and using a Markov model to predict whether the host machine will enter a high load state during the next period of time. However, the authors of the paper only considered the current state of the CPU at the next moment of impact, but did not take into account the fact that the current time before the CPU state will also have an impact on the next moment; therefore, the ordinary single-order Markov chain model will make a very large error in forecasting the host load for some time in the future.

3 System Design

3.1 System Model

During the running of the computer, a portion of the resources (such as CPU, memory, etc.) are used at each moment, resulting in the workload of the computer. On the cloud platform, the physical host on each virtual machine at running time only uses the physical host part of the CPU and RAM and other resources. The physical host workload of a group of virtual machines created after the CPU and other resources is generated by the composition. At the beginning, we assume that the CPU usage of the host measured at a series of discrete time points can be described as a time-separated discrete-time Markov chain.

Assume that $U = u_1, u_2, \ldots, u_n$ is the observed CPU usage history data sequence, the observation interval time t, n is the total amount of current data. The CPU status, $C = c_1, c_2, \ldots, c_m$, m is the number of states divided by the CPU, with

C_l^i representing that the CPU in the state of the moment t_l is c_i. Assume, now that k, when it needs to predict for the moment t_{l-1}. In the state space the K Markov model sequence can be described as shown in Eq. (1).

$$\begin{cases} S = (S^k, c_1), (S^k, c_c), \ldots, (S^k, c_m) \\ S^k = (C_{l-k}^x, C_{l-k}^y, \ldots, C_{l-k}^z), 1 \leq x, y, \ldots, z \leq m \end{cases} \tag{1}$$

It can be seen from the Eq. (1) that the current state space has the number of states Image. The CPU state change sequence can be arbitrarily long, but for CPU data, the number of historical CPU data sequences used to calculate the conditional probability cannot be infinite, and as the sequence grows, the number of states of the Markov model results in exponential growth, which leads to an extremely long running time. Thus, when the values are often selected $K \ll n$, n is the length of the CPU data sequence.

Through the defined state space and historical CPU data, according to the maximum likelihood rule, the calculated Markov transition probability is:

$$P(c_i|S^k) = \frac{Frequency(\langle S^k, c_i \rangle)}{Frequency(S^k)} \tag{2}$$

where $Frequency(S^k)$ represents the occurrences of sequence S^k, whereas $Frequency(\langle S^k, c_i \rangle)$ represents the occurrence of state c_i following S^k. Therefore, through Eq. (2), the transition probability matrix can be calculated, which is a m^k order matrix. In particular, m^k represents when the number of CPU status is m, and the number of the model's state space is m^k.

In the K-Order Markov model it is assumed that the state sequence $(C_{l-k}^1, C_{l-k-1}^1, \ldots, C_{l-1}^1)$ is represented as a state S_0, state sequence $(C_{l-k}^2, C_{l-k-1}^2, \ldots, C_{l-1}^2)$ is represented as $S_1, \ldots$, state sequence $(C_{l-k}^m, C_{l-k-1}^m, \ldots, C_{l-1}^m)$ is represented as S_{m-1}. Then, the probability of the transition from S_0 to S_0 can be calculated through Eq. (3).

$$\begin{aligned} P_{00} &= P\{S_0|S_0\} \\ &= P\left\{\left(C_{l-k-1}^1, C_{l-k-2}^1, \ldots, C_{l-1}^1\right) | \left(C_{l-k}^1, C_{l-k-1}^1, \ldots, C_{l-1}^1\right)\right\} \end{aligned} \tag{3}$$

For the K-Order Markov model, assuming that only n predicts the future load state of the time, the transition probability matrix calculation requires n steps, i.e., P^n. Considering the sequence of Markov with K Image describes the state history K time of the transition probability at that moment, which is therefore to be calculated before the present time point n states space right at the current time weight step reaches $K + n - 1$, i.e., calculates the $K + n - 1$ order Pearson self-correlation coefficient.

The final prediction result is the probability that the CPU will present each state at the next moment, often encountering a situation where the probability of occurrence of two or more states is high and the probability difference is small. If the state with the highest probability of blind selection is selected as the CPU state at the next point in time, the accuracy is not necessarily high. Consider setting a threshold, and when the probabilistic difference between the state with the highest probability and the probability of the probability is greater than this threshold, the CPU will be considered to be in the state at the next moment.

3.2 Algorithm

Order state space K Markov model state space K of the CPU sub-combinations of size K Power CPU state space size, after determining the Markov state sequence space K, calculated based on the historical state of each CPU Markov. The probability of the transition to the other Markov states (including the transfer to their own), after the merger, is a step-by-step transition probability matrix. Calculate the similarity between the CPU states of each step, that is, the Pearson correlation coefficient, and then the CPU state of each step can be determined at the next moment the CPU may be in the various states of the impact of weight. When K is greater than 1, the state of the CPU is affected by the previous K state at the next moment; thus, it is necessary to correct the combined weights of the previously calculated weights as K states. Through the weight and transfer probability matrix, the CPU can be predicted at the next moment in the state of the probability. Similarly, the CPU state of the next time (multiple times) can be predicted. For each time, the forecast value must be tested.

4 Experiment

4.1 Experimental Design

CloudSim is a JAVA package that can be used to simulate cloud computing environments for resource management and scheduling experiments. CloudSim simulates the virtual machine allocation (including CPU, memory, storage space, and bandwidth allocation) in the cloud data center, providing an interface to the upper layer. Users can use the code simulation to create a cloud data center and invoke the corresponding interface to create a virtual machine.

The hardware environment used in our experiments was an Inspur In-Cloud SmartData Appliance (SDA) provided by the Embedded and Pervasive Computing Lab at Huazhong University of Science and Technology. This SDA consists of two main clusters: (1) an admin cluster with two nodes, providing 64 CPU cores, 256 GB of RAM and 3.6 TB of storage, and (2) a worker cluster with seven nodes, providing 84 CPU cores, 336 GB of RAM, and 252 TB of storage.

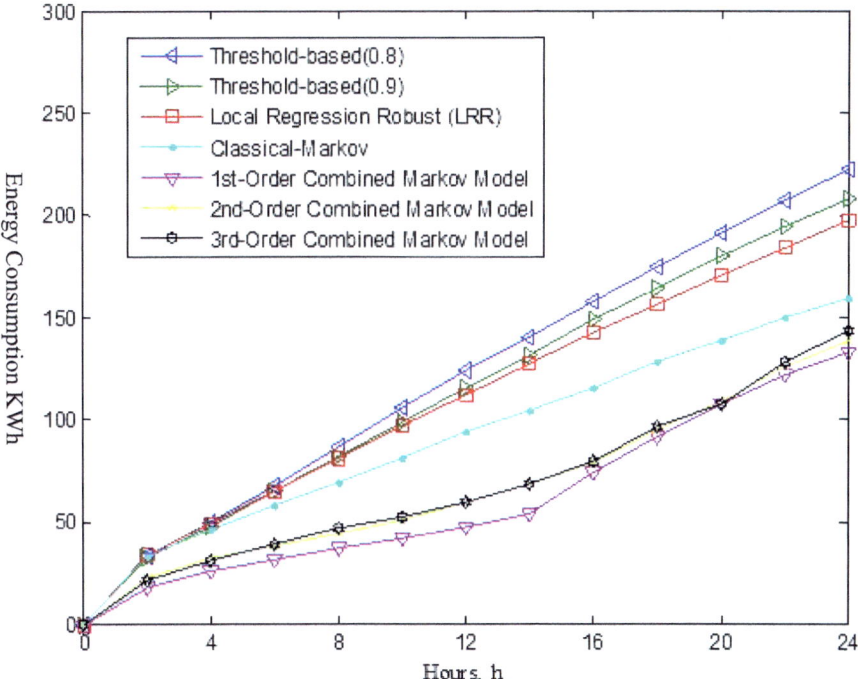

Fig. 2 Comparison of energy consumption

5   Conclusions

This paper studies the timing prediction model, considering the predicted effect of a multi-order Markov model of CPU time and different states affecting the weight and other factors, presents a mixed sequence K Markov model, which also predicts the results of tests to reduce the probability of forecast error as much as possible. By contrast, experiments show that the proposed trigger policy based on the K-sequence hybrid Markov model has great advantages over the traditional model of triggering policy, can effectively reduce the number of virtual machine migrations and cloud data center energy consumption, and will only be lowered to increase the cloud data center system violation of the SLA.

References

1. Lu, H., Li, Y., Zhang, Y., Chen, M., Serikawa, S., & Kim, H. (2017). Underwater optical image processing: A comprehensive review. *Mobile Networks and Applications*. [Online]. Available: https://doi.org/10.1007/s11036-017-0863-4.

2. Li, Y., Lu, H., Li, J., Li, X., Li, Y., & Serikawa, S. (2016). Underwater image de-scattering and classification by deep neural network. *Computers & Electrical Engineering, 54*, 68–77.
3. Jain, R., & Paul, S. (2013). Network virtualization and software defined networking for cloud computing: A survey. *IEEE Communications Magazine,51*(11), 24–31.
4. Huimin, L. U., Yujie, L. I., Nakashima, S., & Serikawa, S. (2016). Turbidity underwater image restoration using spectral properties and light compensation. *IEICE Transactions on Information and Systems, 99*(1), 219–227.
5. Lu, H., Li, Y., Zhang, L., & Serikawa, S. (2015). Contrast enhancement for images in turbid water. *Journal of The Optical Society of America A-optics Image Science and Vision, 32*(5), 886–893.
6. Chen, M., Shi, X., Zhang, Y., Wu, D., & Guizani, M. (2017). Deep features learning for medical image analysis with convolutional autoencoder neural network. *IEEE Transactions on Big Data, PP*(99), 1–1.
7. Li, D., Shang, Y., & Chen, C. (2014). Software defined green data center network with exclusive routing. In *INFOCOM, 2014 Proceedings IEEE* (pp. 1743–1751). Piscataway: IEEE.
8. Zhang, Y., & Min, C. (2016). *Cloud based 5G wireless networks*. Berlin: Springer.
9. Chen, M., Zhang, Y., Hu, L., Taleb, T., & Sheng, Z. (2015). Cloud-based wireless network: Virtualized, reconfigurable, smart wireless network to enable 5G technologies. *Mobile Networks and Applications, 20*(6), 704–712 [Online]. Available: https://doi.org/10.1007/s11036-015-0590-7.
10. Guo, C., Yuan, L., Xiang, D., Dang, Y., Huang, R., Maltz, D., et al. (2015). Pingmesh: A large-scale system for data center network latency measurement and analysis. *ACM SIGCOMM Computer Communication Review, 45*(4), 139–152.
11. Liu, Q., Ma, Y., Alhussein, M., Zhang, Y., & Peng, L. (2016). Green data center with IoT sensing and cloud-assisted smart temperature control system. *Computer Networks, 101*, 104–112.
12. Zhang, Y., Qiu, M., Tsai, C., Hassan, M. M., & A. Alamri. (2017). Health-CPS: Healthcare cyber-physical system assisted by cloud and big data. *IEEE Systems Journal, 11*(1), 88–95.
13. Calheiros, R. N., Ranjan, R., Beloglazov, A., De Rose, C. A., & Buyya, R. (2011). CloudSim: A toolkit for modeling and simulation of cloud computing environments and evaluation of resource provisioning algorithms. *Software: Practice and Experience, 41*(1), 23–50.
14. Al-Ayyoub, M., Jararweh, Y., Daraghmeh, M., & Althebyan, Q. (2015). Multi-agent based dynamic resource provisioning and monitoring for cloud computing systems infrastructure. *Cluster Computing, 18*(2), 919–932.
15. Beloglazov, A., & Buyya, R. (2013). Managing overloaded hosts for dynamic consolidation of virtual machines in cloud data centers under quality of service constraints. *IEEE Transactions on Parallel and Distributed Systems, 24*(7), 1366–1379.
16. Ahmad, R. W., Gani, A., Hamid, S. H. A., Shiraz, M., Yousafzai, A., & Xia, F. (2015). A survey on virtual machine migration and server consolidation frameworks for cloud data centers. *Journal of Network and Computer Applications, 52*, 11–25.
17. Wood, T., Shenoy, P. J., Venkataramani, A., & Yousif, M. S. (2007). Black-box and gray-box strategies for virtual machine migration. In *NSDI'07 Proceedings of the 4th USENIX Conference on Networked Systems Design & Implementation* (Vol. 7, pp. 17–17). Berkeley, CA: USENIX Association.
18. Zhu, X., Young, D., Watson, B. J., Wang, Z., Rolia, J., Singhal, S., et al. (2008). 1000 islands: Integrated capacity and workload management for the next generation data center. In *International Conference on Autonomic Computing, 2008. ICAC'08* (pp. 172–181). Piscataway: IEEE.
19. Gmach, D., Rolia, J., Cherkasova, L., Belrose, G., Turicchi, T., & Kemper, A. (2008). An integrated approach to resource pool management: Policies, efficiency and quality metrics. In *IEEE International Conference on Dependable Systems and Networks With FTCS and DCC, 2008. DSN 2008* (pp. 326–335). Piscataway: IEEE.

20. Gmach, D., Rolia, J., Cherkasova, L., & Kemper, A. (2009). Resource pool management: Reactive versus proactive or let's be friends. *Computer Networks, 53*(17), 2905–2922.
21. Verma, A., Dasgupta, G., Nayak, T. K., De, P., & Kothari, R. (2009). Server workload analysis for power minimization using consolidation. In *Proceedings of the 2009 Conference on USENIX Annual Technical Conference* (pp. 28–28). Berkeley, CA: USENIX Association.
22. Weng, C., Li, M., Wang, Z., & Lu, X. (2009). Automatic performance tuning for the virtualized cluster system. In *29th IEEE International Conference on Distributed Computing Systems, 2009. ICDCS'09* (pp. 183–190). Piscataway: IEEE.
23. Bobroff, N., Kochut, A., & Beaty, K. (2007). Dynamic placement of virtual machines for managing SLA violations. In *10th IFIP/IEEE International Symposium on Integrated Network Management, 2007. IM'07* (pp. 119–128). Piscataway: IEEE.
24. Huang, Q., Shuang, K., Xu, P., Li, J., Liu, X., & Su, S. (2014). Prediction-based dynamic resource scheduling for virtualized cloud systems. *Journal of Networks, 9*(2), 375–383.
25. Park, K., & Pai, V. S. (2006). CoMon: A mostly-scalable monitoring system for planetlab. *ACM SIGOPS Operating Systems Review, 40*(1), 65–74.

Research into the Adaptability Evaluation of the Remote Sensing Image Fusion Method Based on Nearest-Neighbor Diffusion Pan Sharpening

Chunyang Wang, Weikuan Shao, Huimin Lu, Hebing Zhang, Shuangting Wang, and Handong Yue

1 Introduction

Remote sensing image fusion refers to several remote sensing images of the same region, according to a certain rule to extract the information of their respective advantages, and to generate compound multisource remote sensing image technology of synthetic images with new spatial, spectral, and temporal features [1, 2]. Fusion can make full use of complementary information of multiple images to greatly improve image quality and improve the use of images in feature extraction, classification, and target recognition. The main methods used are the intensity–hue–saturation (IHS), Brovey, principal component analysis (PCA), Gram–Schmidt, wavelet, pan-sharpening transform methods, in addition to the smoothing filter-based intensity modulation fusion method and the high pass filtering method.

At present, much research has been carried out into the fusion effect of different fusion methods. For example, Chen et al. [3] compared the IHS transform, wavelet transform, PCA transform, four kinds of image fusion methods, tasseled cap transformation from three aspects of the image brightness information index, spatial information retention index, and spectral information retention index. Wang et al., using the MATLAB system from subjective and objective aspects, compared the fusion effect of three fusion methods: of PCA, IHS, and Brovey [4]. On the other

C. Wang (✉) · W. Shao · H. Zhang · S. Wang · H. Yue
School of Surveying and Land Information Engineering, Henan Polytechnic University, Jiaozuo, China
e-mail: jzitzhb@hpu.edu.cn; wst@hpu.edu.cn

H. Lu
Department of Mechanical and Control Engineering, Kyushu Institute of Technology, Kitakyushu, Japan
e-mail: luhuimin@ieee.org

© Springer Nature Switzerland AG 2020
H. Lu, L. Yujie (eds.), *2nd EAI International Conference on Robotic Sensor Networks*, EAI/Springer Innovations in Communication and Computing,
https://doi.org/10.1007/978-3-030-17763-8_4

hand, Lu et al. (2015–2017) presented underwater image quality assessments that can also be used to evaluate the fusion effect. Some of the underwater image processing methods can apply to the processing of fusion images [5–8]. Chen et al. (2016–2017) proposed the information cognitive system on the 5G-Csys platform [9], data caching and computation offloading in 5G ultra-dense cellular networks [10], and deep feature learning system for image analysis [11]. These methods can be applied for the fusion image processing that with a large amount of data and cognitive information and analysis of fusion images can also be used as a smart assessment system in fusion image effects.

Because few studies have been aimed at the fusion effect evaluation of the nearest-neighbor diffusion pan sharpening method, on this basis, this chapter will apply the new image fusion method based on nearest-neighbor diffusion pan sharpening [12] to compare it with the wavelet transform, PCA transform, Gram–Schmidt transform fusion method, and took a WorldView-2 image as an example to carry out a fusion test, analyzing and evaluating the fusion effect of the four fusion methods from qualitative and quantitative aspects.

2 Fusion Effect Evaluation

2.1 Evaluating the Image Information Index

Information Entropy
Information entropy [13] is an important indicator for measuring the degree of image information richness.

$$H = -\sum_{i=0}^{t-1} P_l \log P_l \tag{1}$$

In the formula, P_l image pixel value is l probability and T is the image gray level; the larger the H value, the greater the richness of information contained in the image.

Standard Deviation
Standard deviation (σ) measures the degree of dispersion of an image's pixel value relative to its mean [14].

$$\sigma = \sqrt{\frac{\sum_{i=1}^{M}\sum_{j=1}^{N}(F(i,j)-\mu)^2}{M \times N}} \tag{2}$$

2.2 Index of Evaluating the Ability of Image Spatial Information Retention

Mean Gradient
The average gradient G reflects the small details of the image contrast and texture transform feature.

$$G = \frac{1}{(M-1) \times (N-1)} \sum_{i=1}^{M-1} \sum_{j=1}^{N-1} \sqrt{\frac{\Delta f_x^2(i,j) + \Delta f_y^2(i,j)}{2}} \tag{3}$$

Spatial Frequency
Spatial frequency (Sf) is a measure of the overall degree of activity in the image space domain. If the Sf is greater—indicating that the higher the degree of activity in the image space domain, the clearer the image—the better the image.

$$R_f = \sqrt{\frac{1}{M \times N} \sum_{i=1}^{M} \sum_{j=2}^{N} [F(i,j) - F(i,j-1)]^2} \tag{4}$$

$$C_f = \sqrt{\frac{1}{M \times N} \sum_{i=2}^{M} \sum_{j=1}^{N} [F(i,j) - F(i-1,j)]^2} \tag{5}$$

2.3 Index of Evaluating Image Spectral Information Retention

Deviation Index
The deviation index (D) reflects the deviation between the fused image and the original image [15].

$$D = \frac{1}{M \times N} \sum_{i=1}^{M} \sum_{j=1}^{N} \frac{\left| I(i,j) - A(i,j) \right|}{A(i,j)} \tag{6}$$

Relative Global Dimension Synthesis Error
The relative global dimension synthesis error (ERGAS) is also an index for evaluating the overall spectral change of the fused image relative to the original multispectral image [16].

$$\text{ERGAS} = 100\frac{h}{l}\sqrt{\frac{1}{n}\sum_{k=1}^{n}\left(\frac{\text{RMSE}_k}{\mu_k}\right)^2} \tag{7}$$

3 Experiment and Analysis

3.1 Experimental Data

WorldView-2 is a commercial satellite remote sensing satellite Digital Globe that was launched on 8 October 2009 at Vandenberg Air Force Base in California, USA. It contains a 0.5-m resolution panchromatic image and a 2.0-m resolution multispectral image. The image used in this chapter is a WorldView-2 panchromatic and multispectral image of a region in the south of China, as shown in Fig. 1.

3.2 Experimental Results and Discussion

The panchromatic image and multispectral image respectively use the wavelet, Gram–Schmidt, PCA, and NNDiffuse, four image fusion methods, to carry out the fusion, finally producing the true color composite image (bands 5, 3, 2) shown in Fig. 2a–d.

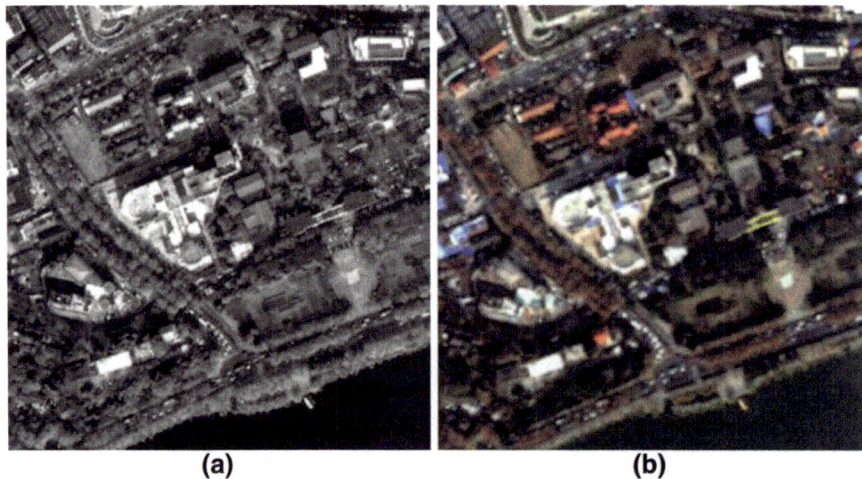

(a)   **(b)**

Fig. 1 Original image data. (**a**)WorldView-2 panchromatic image. (**b**) WorldView-2 true color composite image

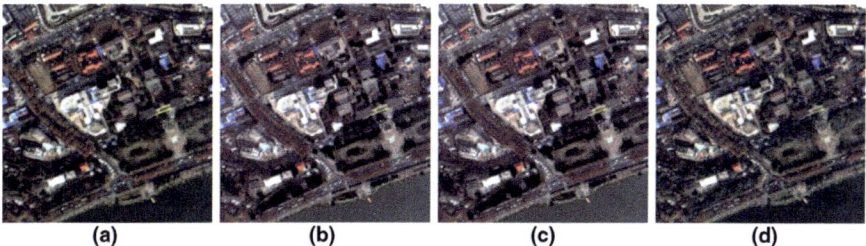

(a)   (b)   (c)   (d)

Fig. 2 The fused image. (**a**) Wavelet fusion method. (**b**) principal component analysis fusion method. (**c**) Gram–Schmidt fusion method. (**d**) NNDiffuse fusion method

Table 1 Comparison of the calculation results of image information

	Panchromatic	Multi-spectrum	Wavelet	PCA	Gram–Schmidt	NNDiffuse
Information entropy	7.3929	6.9433	6.9237	7.0852	7.0180	7.2814
Standard deviation	56.9068	48.2730	47.2342	49.5950	48.1500	56.9258

Table 2 Comparison of the calculation results of the image capability for spatial information retention

	Panchromatic	Multispectrum	Wavelet	PCA	Gram–Schmidt	NNDiffuse
Mean gradient	10.1973	5.3381	5.6344	8.8072	8.9368	9.8629
Spatial frequency	27.0715	18.7034	12.9493	22.1986	22.1970	25.5439

From the overall visual effects, the four fusion methods are different in enhancing image detail information and maintaining color information. In terms of enhanced image detail, NNDiffuse has the best fusion effect, and the object edge information is clear, the contrast is strong, and the visual effect is the best. The Gram–Schmidt and PCA fusion effects are second-best. Wavelet fusion is the worst, as the fusion image is slightly fuzzy. In the respect of preserving color information, the wavelet fused image is close to the original multispectral image display, and there is little change in the Gram–Schmidt and PCA fusion images and the original multispectral image display effect. The color of all kinds of features show certain changes. The NNDiffuse fusion significantly deepened the color of Figs. 1b and 2d, and the original multispectral image of the difference between the color and object contrast (Tables 1 and 2).

It can be seen from Table 3 that the order of increasing information content of the four fusion methods is NNDiffuse > PCA > Gram–Schmidt > wavelet. In terms of image spatial information, the average order of the spatial information-preserving ability of the four fusion methods is NNDiffuse > Gram–Schmidt > PCA > wavelet, which is consistent with the conclusion of the visual judgment.

Table 3 Comparison of the calculation results of image capability for spectral information retention

	Panchromatic	Multispectrum	Wavelet	PCA	Gram–Schmidt	NNDiffuse
Deviation coefficient	–	–	0.0487	0.0511	0.0381	0.0526
Relative global dimension error	–	–	29.3766	41.844	33.5526	38.2412

4 Conclusion

This paper compared the new method NNDiffuse with other frequently used methods, chose the Worldview-2 data for a comparative experiment, and evaluated the effect of the fusion image in both qualitative and quantitative respects. The results show that NNDiffuse is superior to the other three methods in terms of preservation of spatial information, which provides a reference for the future integration of Worldview-2 data and for other data fusion.

Acknowledgments This research is supported by the key research project fund of the Institution of Higher Education in Henan Province (18A420001), the Henan Polytechnic University Doctoral Fund (B2016-13), and The Open Program of the Collaborative Innovation Center of Geo-Information Technology for Smart Central Plains, Henan Province (2016A002).

References

1. Liu, Z., Blasch, E., & John, V. (2017). Statistical comparison of image fusion algorithms: Recommendations. *Information Fusion, 36*, 251–260.
2. Huimin, L., Li, Y., Shota, N., Hyongseop, K., & Seiichi, S. (2013). *Principles and methods of remote sensing application analysis.* Beijing: Science Press.
3. Chen, C., Qin, Q., Wang, J., et al. (2011). Comparison of quality evaluation methods for image fusion of farmland remote sensing. *Transactions of the CSAE, 27*(10), 95–100.
4. Wang, L., Niu, X., Wei, B., et al. (2015). Study on quality evaluation methods for remotely sensed images fusion. *Bulletin of Surveying and Mapping, 2*, 77–79.
5. Li, Y., Lu, H., Li, J., et al. (2016). Underwater image de-scattering and classification by deep neural network. *Computers and Electrical Engineering, 54*, 68–77.
6. Lu, H., Li, Y., Nakashima, S., et al. (2016). Turbidity underwater image restoration using spectral properties and light compensation. *IEICE Transactions on Information and Systems, 99*(1), 219–227.
7. Lu, H., Li, Y., Zhang, L., & Serikawa, S. (2015). Contrast enhancement for images in turbid water. *Journal of the Optical Society of America A, 32*(5), 886–893.
8. Lu, H., Li, Y., Zhang, Y., et al. (2017). Underwater optical image processing: A comprehensive review. *Mobile Networks and Applications, 22*(6), 1204–1211.
9. Chen, M., Hao, Y., Qiu, M., et al. (2016). Mobility-aware caching and computation offloading in 5G ultra-dense cellular networks. *Sensors, 16*, 974.

10. Chen, M., Yang, J., Hao, Y., et al. (2017). A 5G cognitive system for healthcare. *Big Data and Cognitive Computing, 1*, 2. https://doi.org/10.3390/bdcc1010002.
11. Chen, M., Shi, X., Zhang, Y., et al. (2017). Deep feature learning for medical image analysis with convolutional autoencoder neural network. *IEEE Transactions on Big Data.* https://doi.org/10.1109/TBDATA.2017.2717439
12. Sun, W., & Messinger, D. (2014). Nearest-neighbor diffusion-based pan-sharpening algorithm for spectral images. *Optical Engineering, 53*(1), 013107.
13. Shannon, C. E. (2014). A mathematical theory of communication. *Bell System Technical Journal, 27*(3), 379–423.
14. Rodgers, J. L., & Nicewander, W. A. (1988). Thirteen ways to look at the correlation coefficient. *The American Statistician, 42*(1), 59–66.
15. Schwartz, M. H., & Rozumalski, A. (2008). The gait deviation index: A new comprehensive index of gait pathology. *Gait and Posture, 28*(3), 351–357.
16. Bennis, D., Garcia Rozas, J. R., & Oyonarte, L. (2016). Relative Gorenstein global dimension. *International Journal of Algebra and Computation, 26*(8), 1597–1615.

Estimation of Impervious Surface Distribution by Linear Spectral Mixture Analysis: A Case Study in Nantong, China

Ping Duan, Jia Li, Xiu Lu, and Cheng Feng

1 Introduction

The evaluation of environmental impact from impervious surface, as one of the representations of urbanization, has already become a hot topic in the urbanization environment study field [1]. The traditional urban information statistics gathering method is relatively slow and cannot timely and quickly acquire desired data. However, remote-sensing technology can be utilized to quickly and efficiently extract urban surface information [2].

The quick development of remote-sensing technology has resulted in a great-leap-forward in the progress of relevant researches on urban impervious surface. There are already many remote-sensing methods used to analyze impervious surface, including the classification regression tree model, the neural network, and so on. However, due to the existence of abundant mixed pixels, the precision of impervious surface extracted using traditional classification methods is not very high. The spectral mixture analysis model, utilizing the conceptual mode of V-I-S (vegetation–impervious–soil) model [3–9], is one of the hotspot researches on the quantitative analysis of current urban environment remote-sensing.

P. Duan (✉)· J. Li · X. Lu · C. Feng
College of Tourism and Geographic Sciences, Yunnan Normal University, Kunming, Yunnan, China

Key Laboratory of Resources and Environmental Remote Sensing for Universities in Yunnan, Kunming, Yunnan, China

Center for Geospatial Information Engineering and Technology of Yunnan Province, Kunming, Yunnan, China

© Springer Nature Switzerland AG 2020
H. Lu, L. Yujie (eds.), *2nd EAI International Conference on Robotic Sensor Networks*, EAI/Springer Innovations in Communication and Computing,
https://doi.org/10.1007/978-3-030-17763-8_5

This paper takes Nantong in Jiangsu province of China, a small and medium-sized city, as a research area. Based on Landsat 8 remote-sensing images of 2013, a mixed pixel decomposition method is utilized to extract impervious surface information in main urban area of Nantong.

2   Data and Method

2.1   Overview of Study Area and Data Source

The main urban area of Nantong, Jiangsu province, is selected as the study area in this paper (as shown in Fig. 1). Nantong is located in the southeastern part of Jiangsu province, facing the Yellow Sea on the east and leaning against the Yangtze River on the south. It faces Chongming Island of Shanghai, and Suzhou across the river and it is reputed as "Shanghai in the North." In this paper, the main urban area of Nantong is selected as a study area to extract impervious surface data. Landsat 8 OLI-TIRS images acquired on April 14, 2013, are used as the data source. The images have relatively good quality without clouds. The study area only involves images of a scene with track number P119/R38.

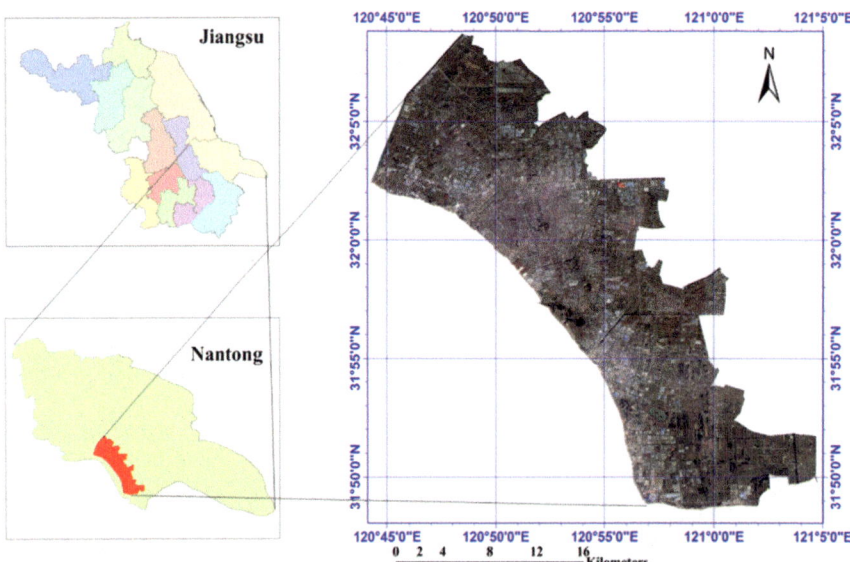

Fig. 1 Location of the study area

2.2 Inversion of Impervious Surface Information

The following steps are usually required to extract impervious surface data by utilizing the linear spectral mixed model [4, 6]: (1) minimum noise fraction (MNF) conversion; (2) manual extraction of end members; (3) collection of terminal land type; (4) decomposition of linear spectral model; and (5) extraction of precision evaluation [10].

2.2.1 Minimum Noise Fraction (MNF)

MNF conversion is a map position space conversion method. MNF conversion is equal to two times the fraction of overlapped principal components. It is an effective method to carry out dimensional reduction of multi-spectral and hyper-spectral remote-sensing images. MNF is sorted according to the size of noise components. After MNF conversion, with the increase of the number of MNF, the amount of information will be gradually reduced and noise will be gradually increased. Table 1 indicates the characteristic value and accumulating contribution rate of each component after MNF conversion.

Through observation of the decomposed seven MNF components, the accumulating contribution rate of characteristic value of the first three MNF components reaches 87%, thus favorably satisfying the demand for extraction of end members. Therefore, only the first three components during extraction of pure pixels occur in the late stage [6].

2.2.2 Extraction of Pixel

The extraction of pixels relies on a two-dimensional scatter diagram comprised of three components of MNF. Figure 2 is the two-dimensional scatter diagram of the first three components after MNF conversion and it presents a general triangular shape. The category of terminal land type is determined according to a category of pixels corresponding to triangular terminal point areas as well as the original standard false color images. These types mainly include vegetation, low albedo (tile, asphalt, etc.), high albedo (concrete, cement, etc.), and land. The spectrum curves of terminal land types extracted are shown in Fig. 3. Since the images are data from Landsat 8, band 5 is near-infrared. Therefore, the reflection DN value of vegetation

Table 1 Characteristic value and cumulative contribution rate of each component after MNF transform

	MNF1	MNF2	MNF3	MNF4	MNF5	MNF6	MNF7
Characteristic value	67.81	13.13	4.89	4.68	3.53	3.1	1.9
Cumulative contribution rate	0.68	0.82	0.87	0.91	0.95	0.98	1.00

Fig. 2 Two-dimensional scatter plot of the first three components of the MNF transform

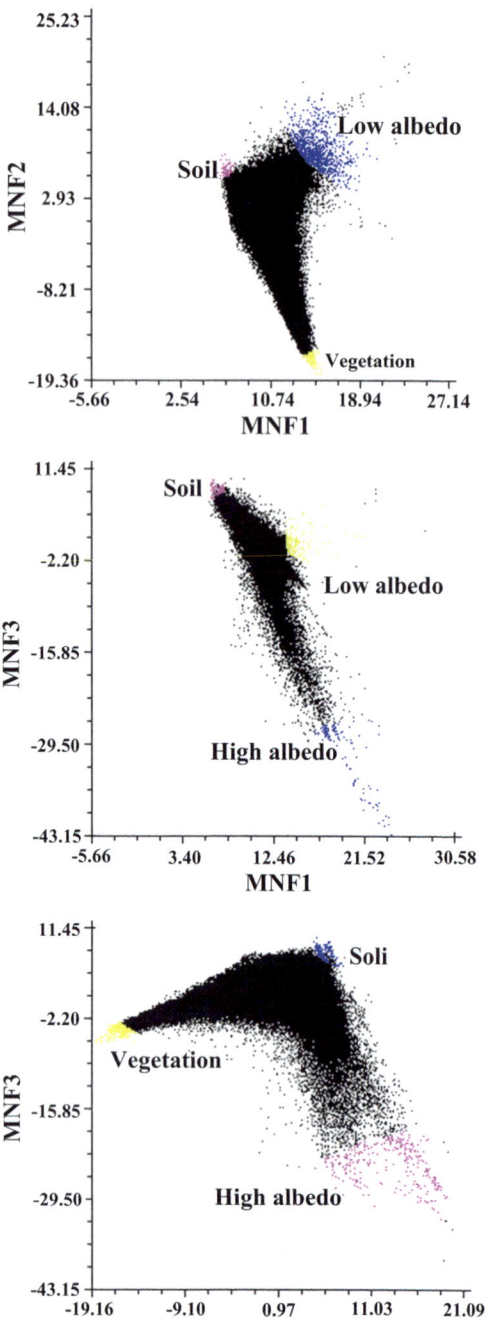

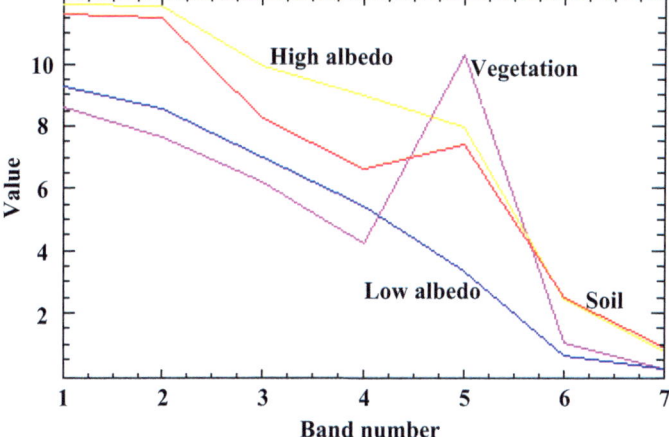

Fig. 3 Spectral profiles of four end members

in band 5 is the highest, while the reflection values of vegetation and soil are between high albedo and low albedo.

3   Results and Discussion

3.1   *Abundant Images of Four Pixels*

Linear spectrum decomposition of Landsat 8 images in the study area is carried out after selection of terminal land type and five data images are obtained. The first four images are high albedo, low albedo, vegetation, and soil data (as shown in Fig. 4).

3.2   *Impervious Surface Percentage (ISP) Distribution Diagram*

In the V-I-S model, impervious surface data was obtained through linear combination of terminal land types with high albedo and low albedo (as shown in Fig. 5). The ISP diagram in Fig. 6 is obtained through ISP calculation of impervious surface. The blue part represents a body of water. The value of ISP gradually increases with the color turning red from green. Surface features can be classified into four types according to empirical thresholds of ISP by referring to the research of Xian & Crane [11] in 2005. The four types of ISP thresholds and main surface features are shown in Table 2. We can see that the value of ISP is very small and is basically

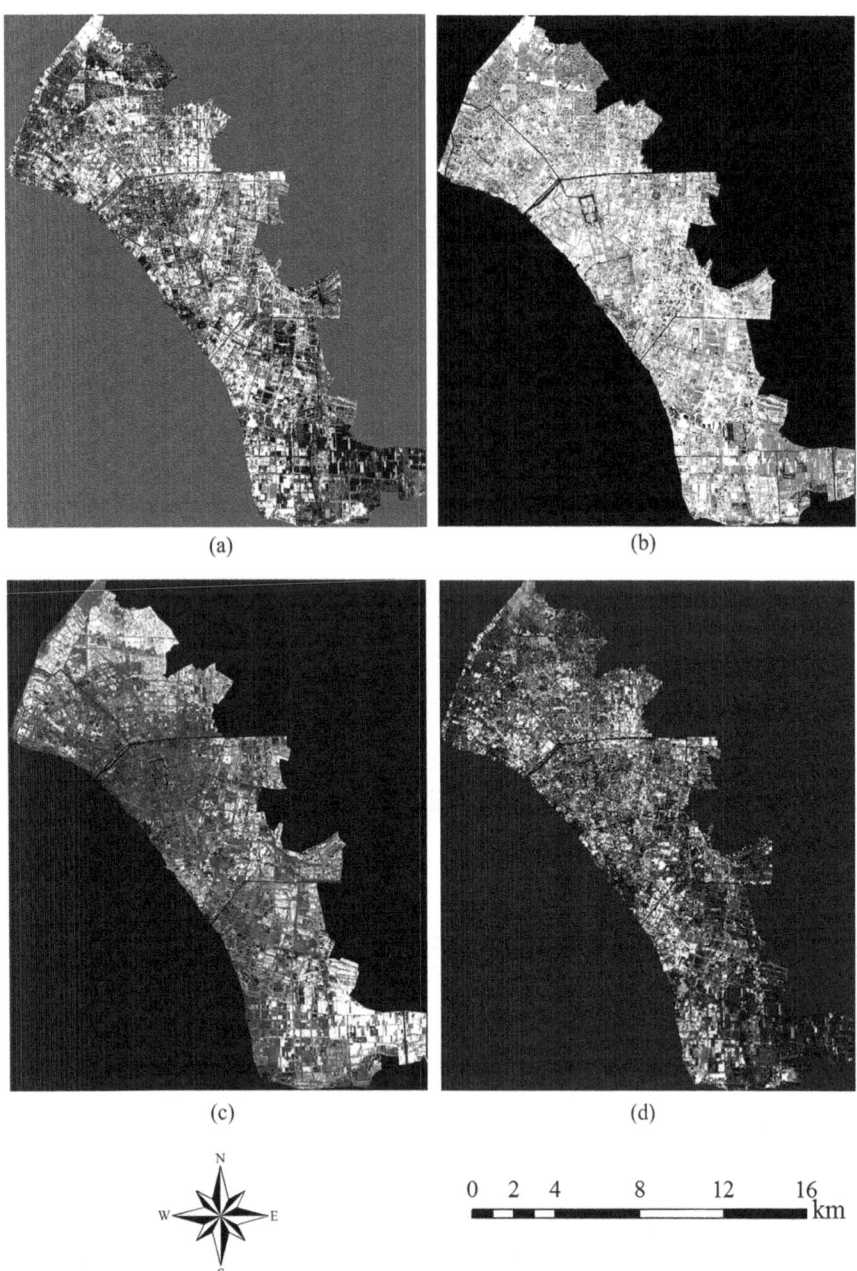

Fig. 4 Abundant images of four pixels with linear spectrum decomposition. (**a**) High albedo. (**b**) Low albedo. (**c**) Vegetation. (**d**) Soil

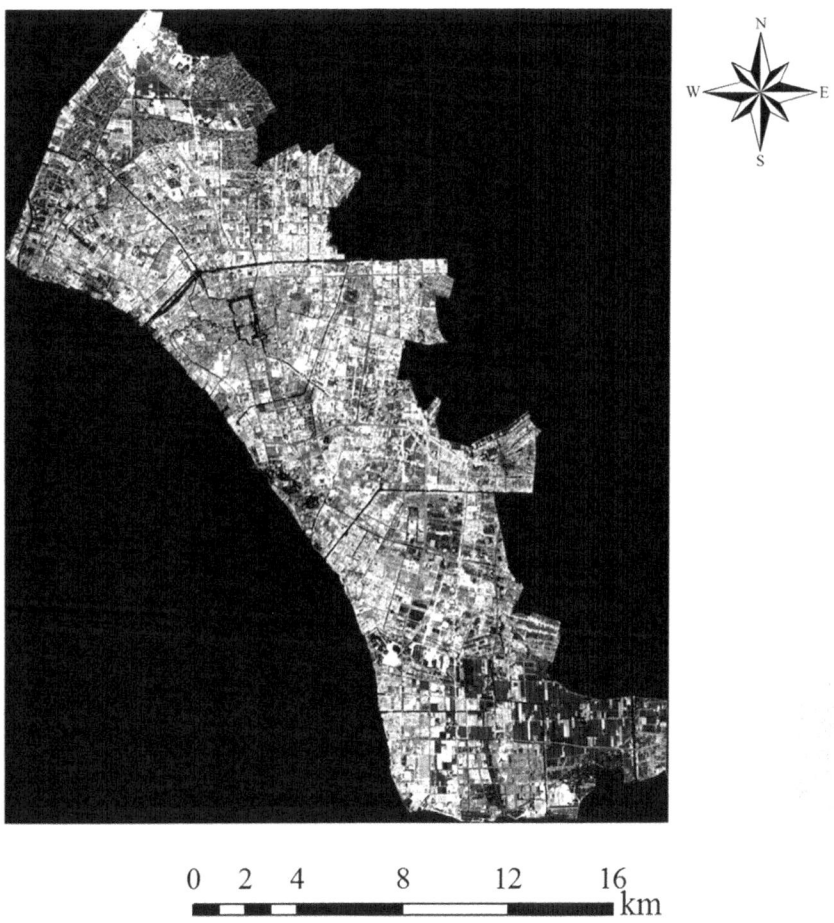

Fig. 5 Fraction image of impervious surface

lower than 30% in agricultural land (cultivated land). The color gradually turns red around urban area. Generally speaking, areas with relatively high ISP value are mainly centralized in central sections of the main urban area. The surrounding regions are main urban areas and are mostly townships and towns with relatively large degrees of cultivated land cover. As a result, the ISP is relatively low.

3.3 Precision Verification

There are two main reasons that result in errors of linear spectral unmixing. One is improper selection of the number of end members (some representative surface

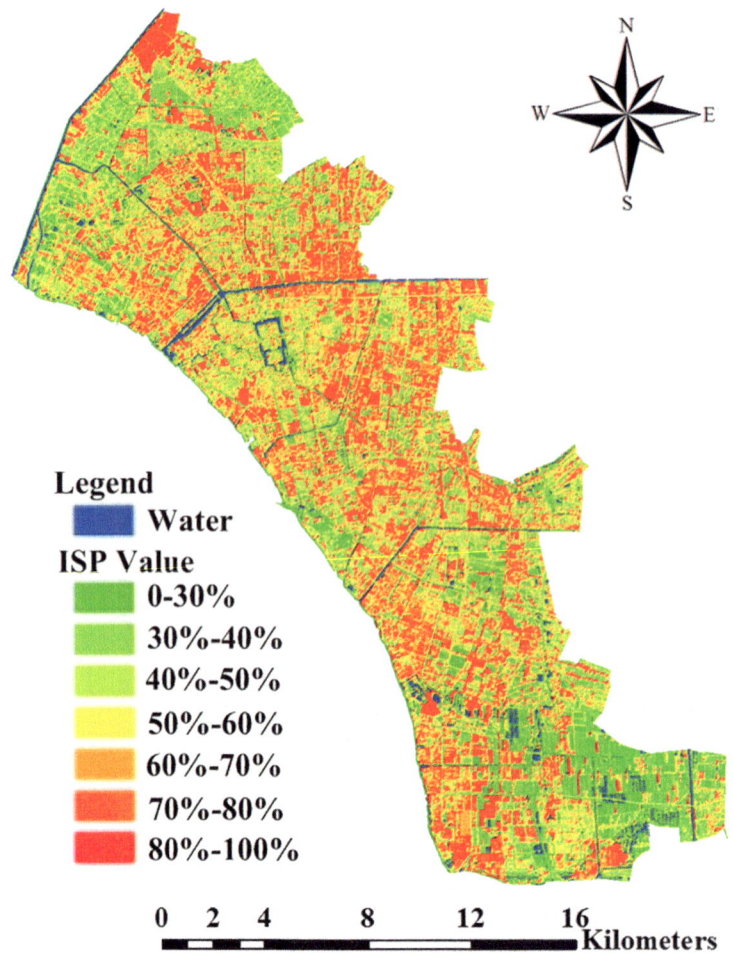

Fig. 6 ISP image of Nantong city

Table 2 The types of urban land use based on ISP empirical threshold

Urban land use type	Non-construction land	Medium and low density urban land	Medium and high density urban land	High density urban land
ISP value/%	<30	31–50	51–80	81–100
Main components	Agricultural land and urban greening, etc.	Residential land and a small number of roads, etc.	Commercial land, old city, roads, etc.	Industrial storage land, commercial land, etc.

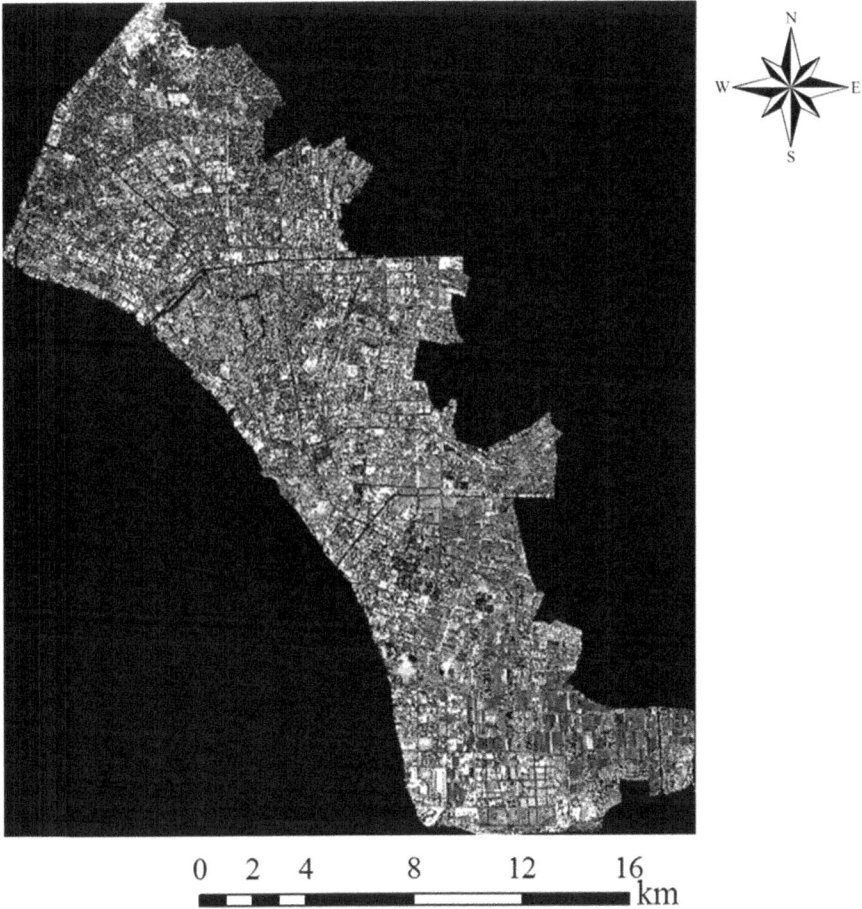

Fig. 7 Root mean square image

feature types would be omitted), and the other is improper selection of the end member spectrum (some spectrum end members selected are not representative). The result of root mean square (RMS) can give a good indication of the accuracy of linear spectral unmixing. Under general circumstances, the mean of decomposition results must be <0.02 [12]. It is concluded from the RMS diagram of the fifth piece of data finally decomposed through extraction of end members and terminal land types (as shown in Fig. 7) that the maximum and the minimum of RMS are 1.054138 and 0.000000, respectively, with mean of 0.002637. In order to meet the requirements for precision and obtain reliable decomposition results, relevant subsequent research can be conducted.

4 Conclusions

In this paper, a spectral mixture analysis method is mainly adopted to decompose and extract impervious surface data. ISP is adopted to analyze the distribution of impervious surface values in the main urban areas of Nantong. The analysis mentioned above indicates that the closer to commercial areas and more densely populated places, the bigger the ISP value will become. The research area of this paper belongs to the plain area and is flat terrain. For non-vegetated areas, especially in the plateau and mountainous areas, there will be serious vegetation shadows. The accuracy of the mixed pixel decomposition is thus affected. Whether the method is applicable for non-urban areas (such as plateau mountains, etc.) remains to be further studied. Generally speaking, the study of this paper can draw the following conclusions:

1. The utilization of a linear spectral analysis model to extract impervious surface data can realize relatively good precision and then the subsequent correlation analysis and ISP analysis among elements will be available.
2. The coverage of impervious surface is higher when the residential areas centering on central sections are dense and there is relatively abundant commercial land based on distribution conditions of ISP values. On the contrary, the coverage of impervious surface is lower.

References

1. Weng, Q. H., & Lu, D. S. (2009). Landscape as a continuum: An examination of the urban landscape structures and dynamics of Indianapolis City, 1991-2000, by using satellite images. *International Journal of Remote Sensing, 30*(10), 2547–2577.
2. Slonecker, E. T., Jennings, D. B., & Garofalo, D. (2001). Remote sensing of impervious surfaces: A review. *Remote Sensing Reviews, 20*(3), 227–255.
3. Ridd, M. K. (1995). Exploring a V-I-S (vegetation-impervious surface-soil) model for urban ecosystem analysis through remote sensing. *International Journal of Remote Sensing, 16*(12), 2165–2185.
4. Voorde, T. V. D., Jacquet, W., & Canters, F. (2011). Mapping form and function in urban areas: An approach based on urban metrics and continuous impervious surface data. *Landscape & Urban Planning, 102*(3), 143–155.
5. Deng, Y., Fan, F., & Chen, R. (2012). Extraction and analysis of impervious surfaces based on a spectral un-mixing method using Pearl River Delta of China Landsat TM/ETM+ imagery from 1998 to 2008. *Sensors, 12*(2), 1846–1862.
6. Lu, D., & Weng, Q. (2006). Use of impervious surface in urban land-use classification. *Remote Sensing of Environment, 102*(1), 146–160.
7. Tang, F., & Xu, H. (2017). Impervious surface information extraction based on hyperspectral remote sensing imagery. *Remote Sensing, 9*(6), 550.
8. Zhu, H., Ying, L. I., & Liu, Z. (2014). Estimation of impervious surface based on semi-constrained spectral mixture analysis. *Remote Sensing for Land & Resources, 26*(2), 48–53.
9. Shen, Y., Shen, H., & Li, H. (2016). Long-term urban impervious surface monitoring using spectral mixture analysis: A case study of Wuhan city in China. In *2016 IEEE International Geoscience and Remote Sensing Symposium (IGARSS)* (pp. 6754–6757). IEEE.

10. Boardman, J. W., & Kruse, F. A. (1994). Automated spectral analysis: A geological example using AVIRIS data, North Grapevine Mountain, Nevada. In *Proceedings of the Thematic Conference on Geologic Remote Sensing* (pp. 407–418). Environmental Research Institute of Michigan.
11. Xian, G., & Crane, M. (2005). Assessments of urban growth in the Tampa Bay watershed using remote sensing data. *Remote Sensing of Environment, 97*(2), 203–215.
12. Wu, C., & Murray, A. T. (2003). Estimating impervious surface distribution by spectral mixture analysis. *Remote Sensing of Environment, 84*(4), 493–505.

Marine Organisms Tracking and Recognizing Using YOLO

Tomoki Uemura, Huimin Lu, and Hyoungseop Kim

1 Introduction

The ocean gives us a lot of blessings that are natural resources, seafood, and so on. However, it is very hard and dangerous for us to investigate the sea. So we hope that the robot takes on underwater tasks because it is too dangerous for us. Thus, "Autonomous Underwater Vehicle: AUV" developed actively. AUV takes on underwater tasks, and it does those tasks automatically. If AUV can automatically investigate the ocean, it would give us useful information to analyze an ecological system, natural resources, and so on. However, a system that investigates deep sea automatically has never developed. A purpose of our research is developing a system for AUV that is possible to recognize a surrounding environment and tracking marine organisms. This system is required a skill of detection and recognition of marine organisms from visual information. We employed a technique of recognition and tracking of multi-objects, which called "You Only Look Once: YOLO" [1, 2]. This method provides us very fast and accurate tracker. Hence we integrated "YOLO" with our system, and calculated the performance of it.

2 Dataset

The dataset we used was provided by the Japan Agency for Marine-Earth Science and Technology (JAMSTEC), which is one of the largest agencies for marine

T. Uemura · H. Lu (✉) · H. Kim
Department of Mechanical and Control Engineering, Kyushu Institute of Technology, Kitakyushu, Japan
e-mail: luhuimin@ieee.org

© Springer Nature Switzerland AG 2020
H. Lu, L. Yujie (eds.), *2nd EAI International Conference on Robotic Sensor Networks*, EAI/Springer Innovations in Communication and Computing, https://doi.org/10.1007/978-3-030-17763-8_6

research in the world. The database of the JAMSTEC E-library of Deep-sea Images (J-EDI) contains approximately 1,300,000 labeled images and 32,000 h of deep-sea videos. In this study, we constructed the first marine organism database, Kyutech10K, with seven categories (i.e., shrimp, squid, crab, shark, sea urchin, manganese, and sand), 10,728 images, and 1489 videos. The images and videos are fixed to a size of 480 × 640 pixels. And, we applied proposed system to four videos (shrimp: 1, squid: 1, crab: 1, shark: 1) in Kyutech10K [3, 4].

3   System Configuration

At first, we remove the scatters, which is caused by turbidity of water, from image. Then, we apply "YOLO" to tracking and recognizing of marine organism.

3.1   Haze Removal on Deep-Sea Images

Underwater imaging models generally follow a standard attenuation model to accommodate wavelength attenuation coefficients. In the proposed method, the Koschmieder model [5–9] is adopted, which estimated as a description of the atmospheric effects of weather on the observer. However, for underwater imaging, the observed irradiance is a linear combination attenuated in the route of sight and the scattered ambient light. Therefore, a modified Koschmieder model is employed for underwater lighting conditions.

The modified Koschmieder model can be expressed as follows:

$$I^c(x) = J^c(x)e^{-\eta d(x)} + \rho(x) \cdot J^c(x)\left(1 - e^{-\eta d(x)}\right), c \in \{r, g, b\} \tag{1}$$

where $J(x)$ is the real scene at depth $D(x)$, $\rho(x)$ is the normalized radiance of a scene point, d is the distance from the scene point to the camera, and η is the total beam attenuation coefficient which is nonlinear and dependent on the wavelength.

The proposed method is based on [7]. We found that turbid underwater images mostly exhibit dark qualities. The minimum operation is suitable for reducing the halo effect when estimating the coarse transmission. Thus, the underwater minimum dark channel priors can be defined as

$$\tilde{d}(x) = \min_{\Re(m,n)} \left(\min_{c \in \{r,b\}} \frac{I^c(x)}{A^c}\right), c \in \{r, b\} \tag{2}$$

where $\Re$ is a square window of size 5 × 5. For each pixel located at position (m, n) in the square patch $\Re$, the values from the red and blue channels are compared, and the lower value is selected. The proposed method can prevent the halo effect around occlusion boundaries. Accordingly, the coarse estimate of transmission is obtained as follows:

$$d(x) = 1 - \omega \tilde{d}(x) \tag{3}$$

where $\omega = 0.8$ for most scenes.

In the above, we discussed the rough estimation of the coarse depth map. However, its depth map contains mosaic effects and yields less accurate results. Therefore, we have developed a joint filter to reduce such mosaic effects. The normalized image is obtained as follows:

$$I_f^c(x) = \begin{cases} \frac{I^c(x) - I_{\min}^c(x)}{I_{\max}^c(x) - I_{\min}^c(x)} & , \text{if } 0 < I^c(x) < 1 \\ 0 & , \text{if } 0 > I^c(x), \quad c \in \{r, g, b\} \\ 1 & , \text{if } 1 < I^c(x) \end{cases} \tag{4}$$

The refinement of the joint filter is first performed under the guidance image $I_f^c(x)$. Here, let $d_p(x)$, $d_q(x)$, $I_{f,p}^c(x)$, and $I_{f,q}^c(x)$ be the intensity value at the pixel p, q of the depth map and the guidance image, respectively, while w_k is the kernel window centered at pixel k. The refined depth map is then formulated as

$$R(x) = \frac{1}{\sum_{q \in w_k} W_{pq}\left(I_f^c(x)\right)} \sum_{q \in w_k} W_{pq}\left(I_f^c(x)\right) d_q(x) \tag{5}$$

where the kernel weight function $W_{G_{pq}}\left(I_f^c(x)\right)$ is expressed as

$$W_{pq}\left(I_f^c(x)\right) = \frac{1}{|w|^2} \sum_{k:(p,q) \in w_k} \left(1 + \frac{\left(I_{f,p}^c(x) - \mu_k\right)\left(I_{f,q}^c(x) - \mu_k\right)}{\sigma_k^2 + \epsilon}\right) \tag{6}$$

where μ_k and σ_k^2 are the mean and variance of the guidance image in the local window w_k, and $|w|$ is the number of pixels in this window. After the refined depth map is obtained, we can recover the real scene using the underwater dark channel prior descattering model.

3.2 Detecting and Tracking Marine Organisms

We detect marine organisms on deep-sea images by using "YOLO." This method is based on convolutional neural networks (CNN) and possible to detect and track multi-objects on image fast. "YOLO" is designed to enable end-to-end learning, in other words it is possible to implement proposing of object regions and predicting of object labels at the same time. The model is shown in Fig. 1 and we describe the outline of "YOLO" below.

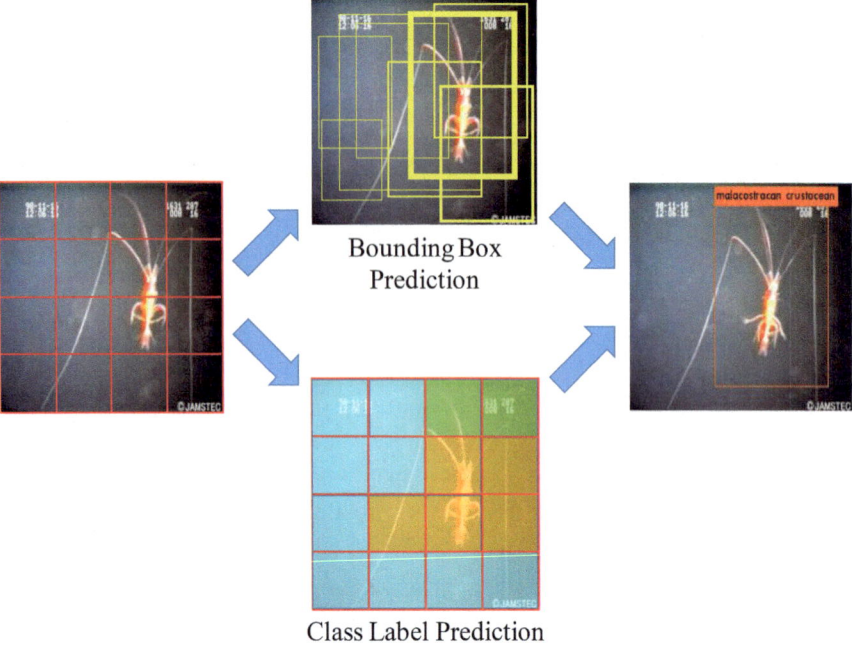

Fig. 1 The model of "YOLO"

"YOLO" divides the input image into an $S \times S$ grid. If the center of an object falls into a grid cell, that grid cell is responsible for detecting that object. Each grid cell predicts B bounding boxes and confidence score: Pr(Object) for those boxes. The confidence score means likelihood that an object exists. More precisely, if an object exists in predicted bounding box, the confidence score should be highest. Each grid cell predicts conditional probability: Pr(Class$_i$) of each class labels too. The highest object label of probability means that object exists in a bounding box. Finally the confidence score of each bounding box: P is determined by the following equation:

$$P = \max \left(\text{Pr} \left(\text{Object} \right) * \text{Pr} \left(\text{Class}_i \right) \right) \tag{7}$$

We select bounding boxes whose confidence score: P exceeds threshold T. In our system, we used $S = 17$ and $B = 3$.

4 Experimental Results and Discussions

We applied our proposed system to four videos. Figure 2 shows an example of detecting results. The threshold $T = 0.24$ is being set for default value in "YOLO." We evaluated the precision and recall on different two thresholds, $T = 0.24$ and

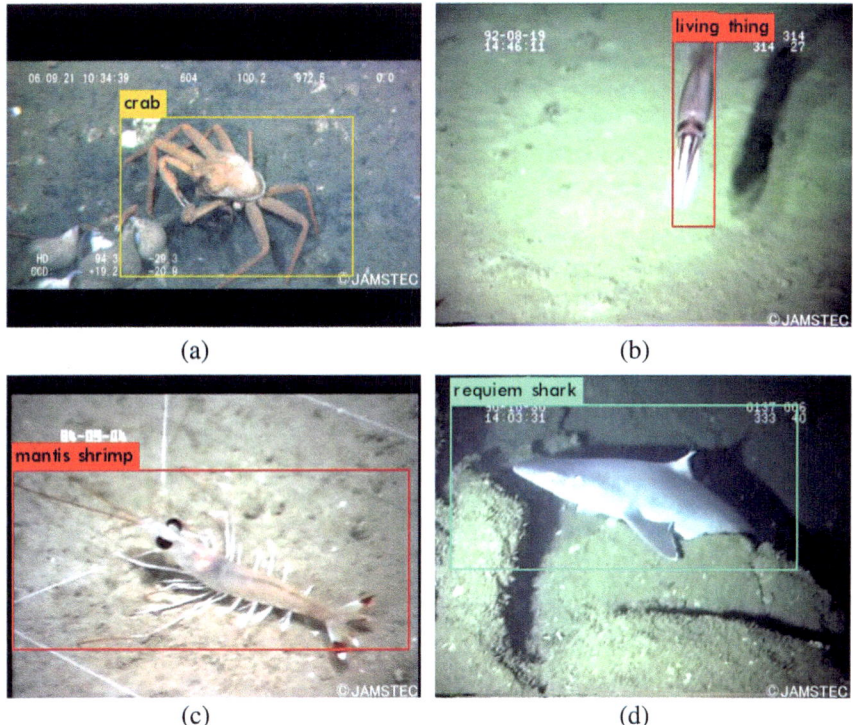

(a)   (b)

(c)   (d)

Fig. 2 An example of detecting results. (**a**) Crab, (**b**) Squid, (**c**) Shrimp, (**d**) Shark

Threshold	Subject	Precision [%]	Recall [%]
0.24	Crab	100 (40/40)	78.4 (40/51)
	Squid	100 (7/7)	8.33 (7/84)
	Shrimp	96.2 (25/26)	24.8 (26/105)
	Shark	100 (6/36)	16.7 (6/36)
0.10	Crab	100 (51/51)	100 (51/51)
	Squid	100 (50/50)	59.5 (50/84)
	Shrimp	85.9 (67/78)	74.3 (78/105)
	Shark	91.3 (21/23)	63.9 (23/36)

Table 1 The precision and recall of each subject on different two thresholds, $T = 0.24$ and $T = 0.1$

$T = 0.1$. The result is shown in Table 1. It is easy to understand that when we use $T = 0.1$, the recall of system is better than $T = 0.24$. However, in general, the precision decreases with decreasing the threshold. So, we have to analyze the characteristics of system, and select an optimum threshold to get best performance.

While tracking of marine organisms shows high performance, prediction of the label is low. For example, most of the shrimps were classified as centipede. The CNN model of "YOLO" is trained by using "ImageNet." "ImageNet" has a large amount of annotated type of images. However, it does not considered the situation that is classification task for marine organisms on deep-sea images. Therefore, we have to retrain the CNN model with deep-sea images.

5 Conclusion

We developed a system which can track and recognize the marine organisms from captured image. The developed system shows generally satisfactory performance. In future, we will use the reinforcement learning methods to improve the classification accuracy [10, 11].

Acknowledgments This work was supported by JSPS KAKENHI (17K14694), Leading Initiative for Excellent Young Researcher (LEADER) of Ministry of Education, Culture, Sports, Science, and Technology—Japan (16809746), Research Fund of State Key Laboratory of Marine Geology in Tongji University (MGK1803), Research Fund of State Key Laboratory of Ocean Engineering in Shanghai Jiaotong University (1510), Research Fund of The Telecommunications Advancement Foundation, Fundamental Research Developing, Association for Shipbuilding and Offshore and Strengthening Research Support Project of Kyushu Institute of Technology. We also thank JAMSTEC for offering the datasets.

References

1. Redmon, J., Divvala, S. K., Girshick, R. B., & Farhadi, A. (2016). You only look once: Unified, real-time object detection. In *2016 IEEE Conference on Computer Vision and Pattern Recognition (CVPR)* (pp. 779–788).
2. Redmon, J., & Farhadi, A. (2017). YOLO9000: Better, faster, stronger. In *2017 IEEE Conference on Computer Vision and Pattern Recognition (CVPR)* (pp. 6517–6525).
3. Li, Y., Lu, H., Li, J., Li, X., Li, Y., & Serikawa, S. (2016). Underwater image de-scattering and classification by deep neural network. *Computers and Electrical Engineering, 54*, 68–77.
4. Lu, H., Li, Y., Uemura, T., Ge, Z., Xu, X., He, L., Serikawa, S., & Kim, H. (2017). FDCNet: Filtering deep convolutional network for marine organism classification. *Multimedia Tools and Applications, 77*, 21847–21860.
5. Lu, H., Li, Y., Mu, S., Wang, D., Kim, H., & Serikawa, S. (2017). Motor anomaly detection for unmanned aerial vehicles using reinforcement learning. *IEEE Internet of Things, 5*(4), 2315–2322.
6. Lu, H., Li, Y., Chen, M., Kim, H., & Serikawa, S. (2018). Brain intelligence: Go beyond artificial intelligence. *Mobile Networks and Application, 23*(2), 368–375.
7. Lu, H., Li, Y., Nakashima, S., & Serikawa, S. (2016). Turbidity underwater image restoration using spectral properties and light compensation. *IEICE Transactions on Information and Systems, E-99D*(1), 219–226.
8. Lu, H., Li, Y., Zhang, L., & Serikawa, S. (2015). Contrast enhancement for images in turbid water. *Journal of Optical Society of America A, 32*(5), 886–893.
9. Serikawa, S., & Lu, H. (2014). Underwater image dehazing using joint trilateral filter. *Computers and Electrical Engineering, 40*(1), 41–50.
10. Chen, M., Yang, J., Hao, Y., Mao, S., & Hwang, K. (2017). A 5G cognitive system for healthcare. *Big Data and Cognitive Computing, 1*(1), 2. https://doi.org/10.3390/bdcc1010002.
11. Chen, M., Shi, X., Zhang, Y., Wu, D., & Guizani, M. (2017). Deep features learning for medical image analysis with convolutional autoencoder neural network. *IEEE Transactions on Big Data*. https://doi.org/10.1109/TBDATA.2017.2717439.

Group Recommendation Robotics Based on External Social-Trust Networks

Guang Fang, Lei Su, Di Jiang, and Liping Wu

1 Introduction

In recent years, research into recommendation robotics has developed very rapidly, and many types have appeared. For example, mobile recommendation robotics, context-aware recommendation robotics, social network recommendation robotics, etc. However, most of the current recommendation robotics can only work for a single user. In fact, many daily activities are carried out by a crowd of people, such as watching movies or TV programs, going to a restaurant for a meal, traveling, and getting service in public. Therefore, it is necessary for recommendation robotics to consider suggestions of a group of people, which is called group recommendation robotics (GRR) [9].

In GRR, group members' preferences may be similar or different. How to get the common preference of group members, alleviate the conflict among group members, and make the recommendation results meet the needs of all group members as far as possible are the key problems to be addressed [8, 13].

In the current social networks, GRR take into account both the strength of the relationship between the members of the group [20, 22, 23] and the influence of social network information on each group member [4, 6, 21], and finally generate group recommendations through aggregation strategies. At present, the main influence of social networking is the social impact of the group members on the GRR.

The structure of this chapter is as follows. In Sect. 2, we introduce the related work recommended by the social networks, and Sect. 3 elaborates on the method based on external trust social networks proposed in this chapter. In Sect. 4 the

G. Fang · L. Su (✉) · D. Jiang · L. Wu
Kunming University of Science and Technology, Kunming, China

© Springer Nature Switzerland AG 2020   59
H. Lu, L. Yujie (eds.), *2nd EAI International Conference on Robotic Sensor Networks*, EAI/Springer Innovations in Communication and Computing,
https://doi.org/10.1007/978-3-030-17763-8_7

experimental results are introduced, and Sect. 5 summarizes the full text and discusses future work.

2 Related Work

2.1 Group Recommendation Robotics

Group recommendation robotics [13] usually generate group preferences by aggregating an individual's ratings. According to Jameson and Smyth [9], the main approaches to generating the preference aggregation are the merging of recommendations made for individuals, the aggregation of ratings for individuals, and the construction of a group preference model. Masthoff [28] presents a compilation of the most important preference aggregation techniques. These basic approaches merge the ratings predicted individually for each item to calculate a global prediction for the group. The selection of a proper aggregation strategy is a key element in the success of a recommendation. The work by Masthoff, as cited in [24] describes a series of experiments that were conducted with real users to determine which strategy performs best. These experiments show that the average and the average without misery strategies perform best from the users' point of view because they seem to obtain similar recommendations to those that emerge from an actual discussion in a group of "humans."

Group recommendation robotics can be classified into two main categories [18, 19]: those that perform an aggregation of individuals' preferences to obtain a possible group evaluation for each candidate item; and those that perform an aggregation of individuals' models in a single group model and generate suggestions based on this model. In the first method, an individual-based recommendation robotic is first used to generate predictions for each group member; then, a group consensus function is used to merge the individual predictions and select ones that are most suitable for the whole group. In the second method, a pseudo user profile is generated from all group members, and an individually based recommendation robotic is then used at runtime to generate recommendations for the pseudo user. By considering the recommendations for individual group members and merging [6] them at run-time to generate group recommendations, this GRR architecture can easily accommodate dynamic groups and tailor its recommendations for each specific scenario.

The individual recommender implements the collaborative filtering (CF) algorithm described in Kelleher and Bridge [11] and group recommendation has largely been studied in the context of CF [2, 3, 21]. We have chosen this algorithm because it is widely used to recommend items when the modeling of user preferences is not a valid option (as in most real scenarios [14, 25, 26] and others [30–33]). This algorithm requires users to rate an initial set of items. Then, those ratings are used to estimate the predicted rating for an unrated item.

2.2 Social Network Recommendation Robotic

One of the important influencing factors of the GRR is the social relationship among the group [4, 12, 21, 27]. Research shows that users prefer to accept the recommendation of trusted users rather than anonymous users [22].

Gartrell [6] proposed the merging of social networks into GRR for the first time. The method is called the rule-based group consensus framework, which not only considers the group members "interest, but also describes the group members" weight differently. The system takes social relations, social frequency, and professional level into account.

With the development of the Internet, the social platform has been the focus of researchers as well. HappyMovie [21] is an online GRR on Facebook; moreover, researchers need to handle two tests: one is a personality test, getting users' personalities through the Thomas–Kilmann instrument [10], and personality is divided from selfishness to tolerance, into a total of five levels. Usually, selfish users are not affected by others, tolerant users are easily impacted, and to reach consensus in the group, vulnerable users are made to change their recommendations. The other one is building a user preference model by choosing movies enjoyed by themselves. Then, group suggestions are generated by aggregating individual preferences.

SocialGR [4] is also a real research system. The system takes many aspects of social factors into account; the main consideration is the trust relationship (TR), social similarity (SS), and social centrality (SC). TR reflects the cohesion between two members by analyzing their affective relationship. SS reflects the alikeness between members, i.e., shared activities, likes, friends, or interests. SC reflects members' reputations in the social network. The basic recommendation robotic utilizes a CF recommendation robotic and the group recommender is determined through the maximizing average satisfaction strategy.

Current social networks and group recommendations usually consider the strength of the relationship among the members of the group; the tie among members would affect the results of GRR. Although some reports [4, 6, 20, 22, 23] considered social factors such as the personality, trust, personal expertise, and social status of the members, once the information was acquired, the steps of the social network GRR were as follows: At first, fetching the list of preferences of all individual recommendation robotics; then increasing or decreasing the members' weight via social factors; finally, obtaining the ultimate results of the GRR using aggregation strategies, which consider the individual recommendations and weight. These methods utilize the social factors to a greater or lesser extent; group members' weights are considered in ones. This means that a few of the members would be ignored because of some group members playing a decisive role, leading to the tendency to influence influential members. Therefore, this chapter focuses on the impact of external members on the GRR; the information of external members and internal members is taken as a whole.

Artificial intelligence (AI) is an important technology during daily social life and economic activities, e.g., machine learning, deep learning [15, 29], and reinforcement learning [16]. Lu [17] wanted to employ brain intelligence (BI) to cope with some complicated work generating new ideas about events without having experienced them by using an artificial life with an imagine function.

3 External-Based Social-Trust Networks Group Recommendation Robotics

3.1 GRR Calculation Framework

The current GRR with social networks take into consideration first, the strength of the relationship among the group [20, 22]; second, the influence of social network information on each group member [4, 6, 21]; and finally, some aggregation strategies to generate group recommendations. At present, the main impact of the social network on the GRR is the social influence of the group members. On the one hand, when the group preferences are inconsistent, those systems take care of the preference of the members with more social influence, and take no notice of some of the intentions of the members with little social influence; on the other hand, it is unreasonable that those systems still consider the influence of social networks while the group have reached consensus. The ideas of this chapter include as follows: correcting the preference rating through social-trust networks when the group rating of an item cannot reach consensus, specifically, using the real ratings of external members who are trusted by the group members to correct the prediction rating of the items. When the disagreement of the group is low, namely, the group tendency is to reach consensus, the influence would be reduced on GRR by a social network, so as to dynamically adjust the influence of social networks and reduce the error on GRR. In fact, most people hope to persuade others to follow their advice when faced with the conflict and large degree of disagreement among the group. At that moment, if there is a role that you trusted, that could tell you what is the best choice, perhaps you would change your mind. Watching a movie with your friends, for instance, maybe you did not intend to watch it with them, but someone you trusted would change your mind as he had just watched it and felt good. This scenario shows that correcting the group preference according to new information when the degree of group disagreement is large in GRR, and the group should be able to reach consensus.

At first, obtaining the list of preferences of each group member by CF, and ensuring a special group (by a randomly selected group or fasten group), and getting the results of GRR using aggregation strategies is called classical group prediction. Then, the disagreement of an item i is calculated using the predictions of group members, Disagreement is greatly affected by external trust networks, and vice versa. The frame and formula of our method are Fig. 1 and formula (1).

$$gpred(G, i) = \begin{cases} (1 - \lambda_i) \cdot CGP(G, i) + \lambda_i \cdot OAR(OG, i) \\ CGP(G, i), \text{ if } OAR(OG, i) = \varnothing \\ OAR(OG, i), \text{ if } CGP(G, i) = \varnothing \end{cases} \qquad (1)$$

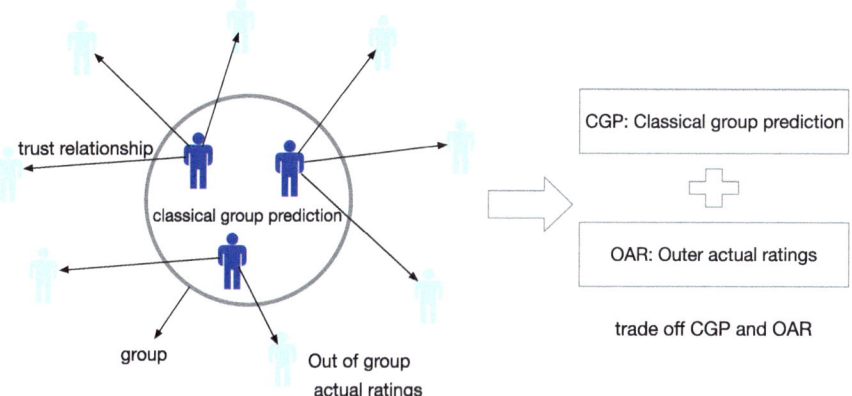

Fig. 1 Frame of group recommendation robotics

Here, G refers to a group, OG (out of group) represents information of the external group, namely users trusted by group members. $CGP(G, i)$ (classical group prediction) is a predictions of item i by individual recommendation robotics and aggregation strategies. $OAR(OG, i)$ (outer actual rating) represents actual ratings of external members of the group, λ_i is a dynamic balance factor of external members and internal members of item i, balancing ratings of recommendations between external and internal members. $\bigsqcup$ represents a aggregation strategy. Note that, if $OAR(OG, i) = \varnothing$, then there are no real ratings of item i by trusted users, and results calculated by $CGP(G, i)$ would be selected. Otherwise, if $CGP(G, i) = \varnothing$, then there are no prediction ratings of item i by group users, and results calculated by $OAR(OG, i)$ would be utilized.

We have adjusted the intensity of the external influence dynamically. When the disagreement of group members is large, we need the external real score to be correct. The greater the disagreement, the greater the external influence of group members. When the disagreement of group members is small, the opinions of the group members need to be retained. The smaller the disagreement, the less the group members would be influenced externally. For example, when two people decide to watch a movie together, even if someone recommends a good movie, it may not be easy to influence the decision of the two people. Two problems need to solved, one is how to determine the value of λ, the other is how to solve $OAR(OG, i)$. $CGP(G, i)$ (classical group prediction) represents predictions of item i by individual recommendation robotics and aggregation strategies. The formula is as follows:

$$CGP(G, i) = \bigsqcup_{\forall u \in G} pred(u, i) \tag{2}$$

$\bigsqcup$ represents a aggregation strategy, $pred(u, i)$ represents user u's prediction rating on item i by individual recommendation robotics.

In $OAR(OG, i)$, the actual ratings of external members, which are relative to the group members, would be introduced into GRR by a social network. Moreover, the object of each user trust may be more than one, and two points need to be considered on external information: degree of trust and obtaining the actual rating set of members who are trusted. The formula is as follows:

$$OAR(OG, i) = \bigsqcup_{\forall u \in G, v \in OG} t_{uv} \cdot \text{actual\_rating}(v, i) \tag{3}$$

Here, G refers to a group, OG (out of group) represents information of the external group, namely, users trusted by the group members, $\bigsqcup$ represents a aggregation strategy, t_{uv} represents the degree of trust of the user u with regard to user v, $t_{uv} \in [0, 1]$, 0 is trust scarcely and 1 means trust completely. The greater the influence, the higher the degree of trust in external information; actualrating(v, i) represents the actual rating of user v on item i.

3.2 Dynamic Adjustment of Parameter λ

LVD This approach utilizes disagreement and refines disagreement a little, first, calculating the unbiased estimate of prediction ratings of group members, namely, sample variance to compute the number of points far away from the center in the group preference set. In the method named lambda via disagreement (LVD), the formula is as follows:

$$\text{dis}(G, i) = \frac{1}{|G| - 1} \sum_{u \in G} \left[pred(u, i) - \frac{\sum_{v \in G} pred(v, i)}{|G|} \right]^2 \tag{4}$$

and the calculating formula of λ is:

$$\lambda = \frac{1}{|G|} \sum_{u \in G} \mathbb{1} \left[\left(pred(u, i) - \frac{\sum_{v \in G} pred(v, i)}{|G|} \right)^2 > \text{dis}(G, i) \right] \tag{5}$$

Here, $|G|$ refers to the number of a group, and an indicator function is introduced: $\mathbb{1}[x]$ refer to that if x is true, then expression is 1, unless 0. $pred(u, i)$ represents user $u's$ prediction rating on item i by individual recommendation robotics.

LVTP Because the structure of LVD is simple, some requirements should be taken on group size and distribution of prediction. On the one hand, the case in only two members cannot be coped with, such as in Table 1, although the disagreements (dis) are 8 and 12.5 respectively, and the values of lambda are both 0, which means that adjustment is unnecessary on recommendations of external social networks, and it is unreasonable. The case with great disagreement is observed; however, as a result

Table 1 λ via disagreement

	User 1	User 2	Disagreements	λ
Item$_A$	0	4	8	0
Item$_B$	0	5	12.5	0

of having small values of lambda, this method cannot deal with the situation of only a few people.

To address the above issue, the LVTP method is proposed in that case consists of the following steps: first, the standard value α_{ri}, a balancing item i, would be set; second, utilizing the standard value α_{ri} divide the predictions of group members into different two parts, a large one is Greater$_{ri}$ and a small one is Less$_{ri}$; finally, the lambda value of each item i is calculated. The formula is as follows:

$$\lambda_i = \frac{\mathbb{1}[\text{Greater}_{ri} \wedge \text{Les}_{ri} > \varnothing](\overline{\text{Greater}_{ri}} - \overline{\text{Les}_{ri}})}{\gamma + \max(\text{Greater}_{ri})} \tag{6}$$

Here, $\overline{\text{Greater}_{ri}}$ represents the average value of a larger than standard value in the rating set. An indicator function is introduced: $\mathbb{1}[x]$ refers to that if x is true, then the expression is 1, unless 0. Note that in step 3, if Greater$_{ri} = \varnothing$ or Less$_{ri} = \varnothing$, which means that the rating of the group of item i reaches consensus and indicates that group members either like or dislike item i, then $\lambda_i = 0$, which means that the social network has almost no influence on item i. Particularly, γ is a smooth factor, to avoid $(\overline{\text{Greater}_{ri}} - \overline{\text{Less}_{ri}}) = \max(\text{Greater}_{ri})$ results in the GRR neglect of the group's suggestions completely, in addition to being unreasonable.

With regard to the choice of standard values, this chapter considers the following three aspects: (1) The mid value of the range can be evaluated, for example, the maximum score for a certain item is 10 and the lowest is 1. The mid-value is $(10 + 1)/2 = 5.5$, and 5.5 is the standard value. (2) The mean of all predictions of item i in the training data is the standard value, and the standard values for each item are different. (3) The median of all predictions of item i in the training data is the standard value, and the standard values for each item are different.

3.3 Group Recommendation Robotics Based on External Social-Trust Networks

A Description of the Aggregation Strategy In GRR, preference fusion refers to integration of the preferences of group members. Preference fusion is also known as the aggregation strategy [24] or the aggregate rules [5]. For consistency of terminology, this chapter uses the term aggregation strategy. Masthoff, cited in [24], described ten aggregation strategies in detail, and the relatively better strategies are thought to be the average strategy and the average without misery strategy, according to a series of experiments in another paper.

In the average strategy, the group rating for a particular item is computed as the average rating over all individuals.

$$gpred(G, i) = \frac{1}{|G|} \sum_{u \in G} pred(u, i) \qquad (7)$$

where $|G|$ is the group size, $pred(u, i)$ is the predicted rating for each user u, and every item i.

Group disagreement with the project [1] $dis(G, i)$ indicates the degree of difference of users in the group G compared with the predicted score of the project i.

$$dis(G, i) = \frac{1}{|G|} \sum_{u \in G} [pred(u, i) - mean(G, i)]^2 \qquad (8)$$

$mean(G, i)$ denotes the mean of the group prediction ratings of item i.

GRITrust Algorithm According to the description of λ and $OAR(OG, i)$ in Sects. 3.1 and 3.2, the result can be computed using those parameters. Note that if there is no prediction on $CGP(G, i)$ but on $OAR(OG, i)$, the GRR result is not empty but $OAR(OG, i)$. The above method can ease the cold-start problem in GRR. Sometimes, the user has no ratings, and the system can help him/her to find better answers.

In this chapter, the algorithm called the group recommender Involve Trust network (GRITrust) pseudo code is as follows.

Algorithm 1 GRITrust

Input: training sets D_{train}, trust network and group G
$D_{train} = \{user_i, item_i, rating_i\}, i \in |D_{train}|$
$Trustnetwork = \{trustor_j, trustee_j, value_j\}$,
$j \in |Trust|, and\ G = \{u_k, u_m, \ldots, u_n\}$
Output: gpred(G)
  Initialization: $pred(u, i) \Leftarrow$ from D_{train} by CF
  repeat
      $CGP(G, i) = \bigsqcup_{\forall u \in G} pred(u, i)$
      $OAR(OG, i) = \bigsqcup_{\forall u \in G, v \in OG} t_{uv} \cdot actual(v, i)$
      $\lambda_i \Leftarrow$ from CGP(G,i) by some methods
      if $CGP(G, i) == \varnothing$ **then**
          gpred(G,i) = OAR(OG,i)
      else $\{OAR(OG, i) == \varnothing\}$
          gpred(G,i)=CGP(G,i)
      else
          $gpred(G, i) = (1 - \lambda_i) \cdot CGP(G, i) + \lambda_i \cdot OAR(OG, i)$
      end if
  until group predictions of all items

---

4 Experiments

4.1 Experimental Data

To verify the proposed method, all performance experiments were conducted on an outline dataset.

Dataset: We have used the FilmTrust [7] ratings dataset for evaluation purposes. This dataset includes 1508 users, 2071 items, and 35,497 ratings whose range is [0.5, 4].

The dataset as follows. First, actual ratings with which users have evaluated movies have been divided into 8 to 2. Then, the training set has 1482 users and the test set has 1421 users, because a few users do not have enough rating counts. In particular, there are only 609 users who can trust one or more other users, which means that just 41% of users would use a social network in the training set. Let p be a ratio value of numbers of users on the network to the training set population, and the value is 0.41. We can calculate the probability of using social networks as $1 - (1 - p)^n$ via binomial distribution when the group size is n, and if we can get $n = 2, 3, 4$, then the probability is 0.65, 0.80, and 0.87 respectively. Obviously, the value is lower than others when n equals 2. Furthermore, to verify the reliability of the proposed method, we would discuss separately the case in which the group size is 2.

In this chapter, a randomized grouping method is used to conduct the experiment. In the group selection, we noticed that if we experiment with a random group of three people, considering each possibility incurs a large computing cost. For example, when there are 1000 users in a dataset, there are 166, 167, 000 randomly selected combinations of three people. Obviously, if the numbers and the sizes of groups are appropriately increased, it is impossible to calculate them one by one. For the sake of convenience, we conducted a random sample of studies.

4.2 Evaluation Method Description

To evaluate the effectiveness of the proposed method, the RMSE [6] is used as an evaluation method that can assess the quality of the recommended system in terms of accuracy. The formula is as follows:

$$\text{RMSE} = \sqrt{\frac{\sum_{i=1}^{N}(gpred(G, i) - \text{actual}(G, i))^2}{N}} \tag{9}$$

where N is the number of items recommended in the group.

4.3 Randomly Divided into Groups Experiment

The basic idea of the experiment is that with use of the CF algorithm in the training set, each user has not seen an item prediction rating, resulting in personalized recommendations. Then, a group is randomly selected and the dynamic adjustment parameters λ are calculated, finding that the group members trusted in the trust network object, selecting the appropriate aggregation strategy and the prediction ratings of the item are calculated according to the formula (1). Finally, comparing predictions with the test set, note that the test set also uses the same aggregation strategy.

The λ_{LVD} in Table 2 indicates the method of calculating λ by LVD and λ_{LVTP} denotes the method of calculating λ by LVTP. Methods of calculating the standard value α are as follows: α_{mean} standard value using the true value of the sample; α_{mid} standard value using the middle of the range can be evaluated; α_{median} standard value using the median of the real sample.

According to the group size, we have made a random selection 100 times in this experiment, and the group size ranges from 3 to 11. Figure 2 illustrates the performance of baseline, GRITrust_dis, GRITrust_mean, GRITrust_mid, GRITrust_median with different group sizes. When the group size is 3, the

Table 2 Explanation

Name	Explanation
Baseline	Without network
GRITrust_dis	λ_{LVD}
GRITrust_mean	$\lambda_{LVTP}\ \alpha_{mean}$
GRITrust_mid	$\lambda_{LVTP}\ \alpha_{mid}$
GRITrust_median	$\lambda_{LVTP}\ \alpha_{median}$

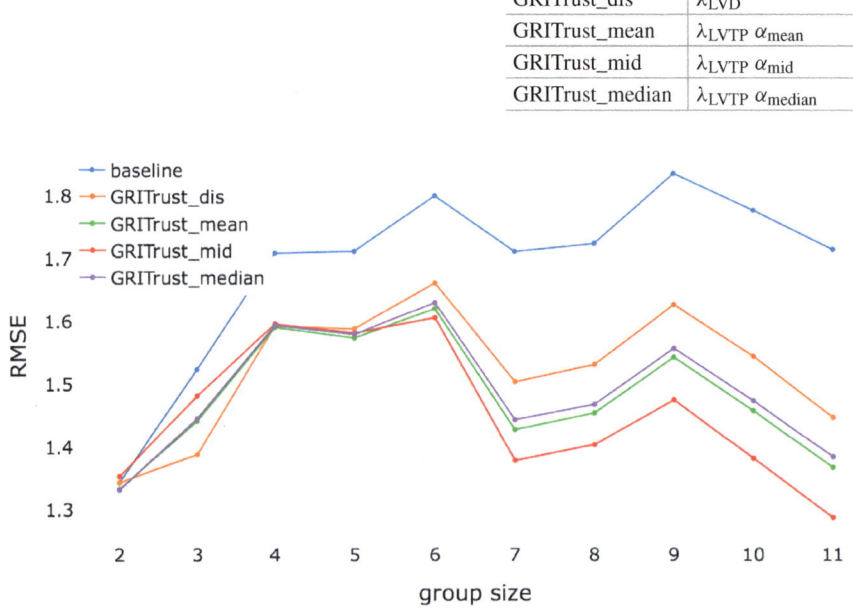

Fig. 2 Root mean square error comparison of different group sizes by the average strategy

GRITrust_dis method is better than the others; this means that LVP is better than LVTP. In addition, if the group size is greater than 5, then LVTP is better than LVP; in particular, GRITrust_mid is better than the others.

4.4 Social-Trust Network Utilization Ratio in Group Recommendations

According to the research, there are three features in GRR with social networks: (1) Not all users are in social networks, which means that some users do not use social networks; (2) In the current social network, some users do not pay attention to others; (3) Some users have little or no rating information on the item. Therefore, we cannot guarantee that users get the useful information every time they visit social networks. The definition of social network utilization is given below.

Definition 1 In GRR with social networks, m is group recommendations, n is social networks are visited and used through recommendation robotics; we define the ratio of n and m as the social network utilization ratio r_{social}.

$$r_{\text{social}} = \frac{n}{m}, \ r_{\text{social}} \in [0, 1] \tag{10}$$

To verify that the proposed method correlates positively with the social network utilization in GRR with social networks, we used the same dataset and method as for the random sample experiment, except that different utilization ratios of social networks were chosen to study the experiment. Among them, $r_{\text{social}} = \{0.2, 0.4, 0.6, 0.8, 1.0\}$, the group size ranges from 3 to 10, for example, G3 indicates that the group size is 3.

In Fig. 3, using the average strategy, we compared RMSE and the social network utilization ratio, which resulted in a positive correlation. Figure 4 shows the percentage fall according to the social network utilization ratio. In Fig. 5, we

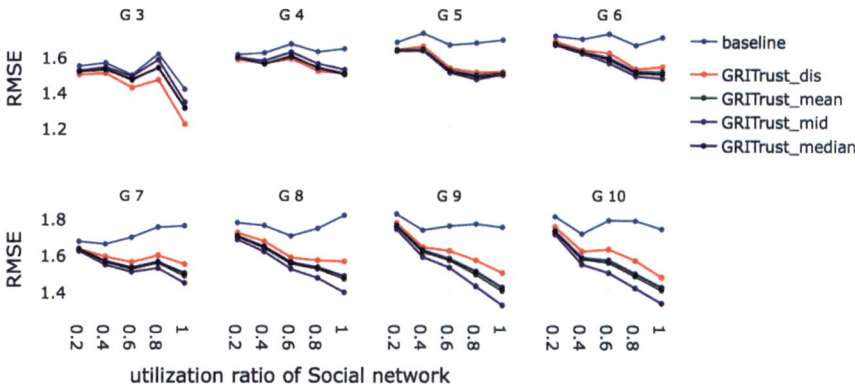

Fig. 3 Root mean square error comparison of different group sizes and different social network utilization ratios

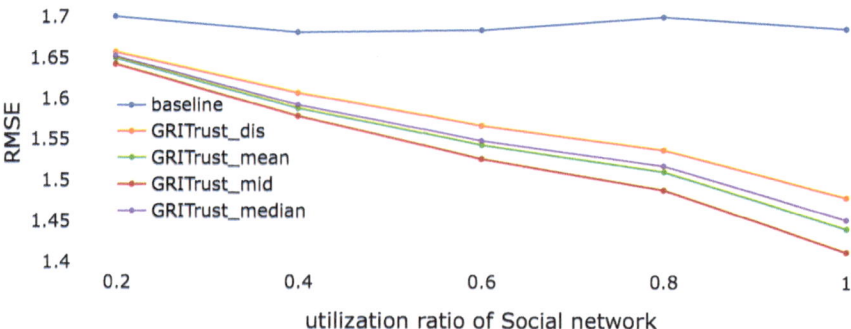

Fig. 4 Root mean square error with utilization of social networks

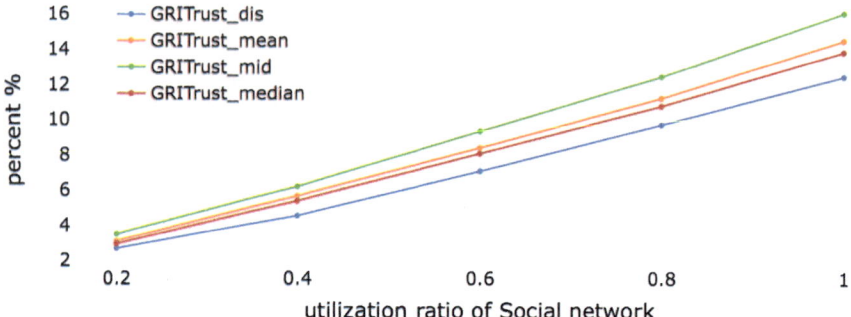

Fig. 5 Percentage fall according to the social network utilization ratio

evaluated RMSE of 20, 40, 60, 80, and 100% of the social network utilization
ratio in different group sizes. The experiment shows that RMSE decreases as the
social network utilization ratio increases, and the effect achieved by GRITrust_mid
is relatively good: about 4–16%.

Figure 6 shows the social network utilization in the case of random sampling
by different group sizes: the larger group, the higher the social network utilization
ratio, and the range is 67–92% in this dataset. According to Fig. 5, the RMSE of our
proposed method decreased by about 9–15%.

5 Conclusion

This chapter introduces the influence of social-trust networks on the group recommendation robotic. In this chapter, we employed social-trust network relationships,
through a true evaluation of an item, to amend the group prediction of an item; when
the group disagreement is small, that is, within the group to achieve the same result,

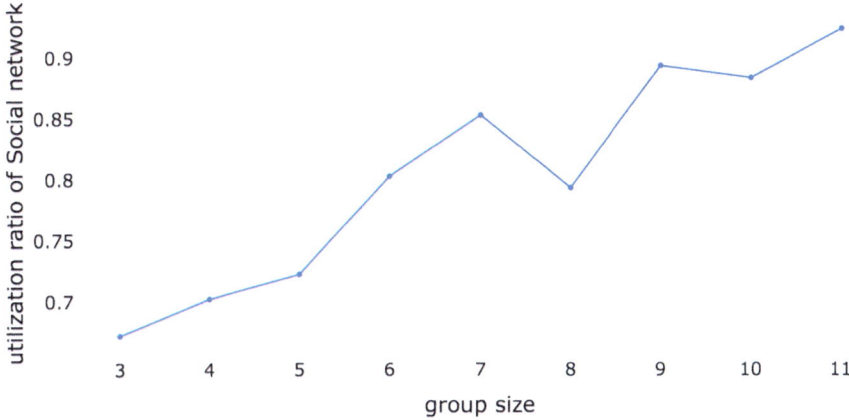

Fig. 6 Random sampling

to reduce the social network recommended to the group impact, thus, dynamically adjusting the impact of social network factors, and improving the quality of group recommendations. Through experiments, we verified the effectiveness of the proposed method. The error of the proposed method did not increase with an increase in the group, and remained at a relatively low level.

Although some results have been achieved, not all the valid information in social networks was fully utilized, such as user similarity, user personality, social status, etc. This chapter was aimed at verifying the impact of group members outside other than discuss the relationship between group members, which will be the future need to be discussed. In addition, other domains need to be considered as well, except for movies, the size of the data is also the direction of efforts.

Acknowledgements This work is supported by the National Science Foundation of China under the Grant No.61365010.

References

1. Amer-Yahia, S., Roy, S. B., Chawlat, A., Das, G., & Yu, C. (2009). Group recommendation: Semantics and efficiency. *Proceedings of the VLDB Endowment, 2*(1), 754–765.
2. Baltrunas, L., Makcinskas, T., & Ricci, F. (2010). Group recommendations with rank aggregation and collaborative filtering. In *ACM Conference on Recommender Systems* (pp. 119–126). New York, NY: ACM.
3. Berkovsky, S., & Freyne, J. (2010). Group-based recipe recommendations: Analysis of data aggregation strategies. In *ACM Conference on Recommender Systems* (pp. 111–118). New York, NY: ACM.
4. Christensen, I. A., & Schiaffino, S. (2014). Social influence in group recommender systems. *Online Information Review, 38*(4), 5–5.

5. Dyer, J. S., & Sarin, R. K. (2011). Group preference aggregation rules based on strength of preference. *Management Science, 25*(9), 822–832.
6. Gartrell, M., Xing, X., Lv, Q., Beach, A., Han, R., Mishra, S., et al. (2010). Enhancing group recommendation by incorporating social relationship interactions. In *International ACM Siggroup Conference on Supporting Group Work, Group 2010, Sanibel Island, Florida, USA, November* (pp. 97–106). New York, NY: ACM.
7. Guo, G., Zhang, J., & Yorke-Smith, N. (2013). A novel Bayesian similarity measure for recommender systems. In *International Joint Conference on Artificial Intelligence* (pp. 2619–2625). Menlo Park, CA: AAAI Press.
8. Jameson, A. (2004). More than the sum of its members: Challenges for group recommender systems. In *Working Conference on Advanced Visual Interfaces* (pp. 48–54). New York, NY: ACM.
9. Jameson, A., & Smyth, B. (2007). Recommendation to groups. In *The Adaptive Web. Lecture Notes in Computer Science* (pp. 596–627). Berlin: Springer.
10. Jones, J. E. (1976). Thomas-Kilmann conflict mode instrument. *Group & Organization Management, 1*(2), 249–251.
11. Kelleher, J., & Bridge, D. (2004). *An accurate and scalable collaborative recommender.* Dordrecht: Kluwer Academic Publishers
12. Kim, J. K., Kim, H. K., Oh, H. Y., & Ryu, Y. U. (2010). A group recommendation system for online communities. *International Journal of Information Management, 30*(3), 212–219.
13. Kompan, M., & Bielikova, M. (2014). Group recommendations: Survey and perspectives. *Computing & Informatics, 33*(2), 1–31.
14. Linden, G., Smith, B., & York, J. (2003). Amazon.com recommendations. *IEEE Internet Computing. January–February, 4*(1), 76–80.
15. Lu, H., Li, B., Zhu, J., Li, Y., Li, Y., Xu, X., et al. (2017). Wound intensity correction and segmentation with convolutional neural networks. *Concurrency & Computation Practice & Experience, 29*(6), e3927.
16. Lu, H., Li, Y., Mu, S., Wang, D., Kim, H., & Serikawa, S. (2017). Motor anomaly detection for unmanned aerial vehicles using reinforcement learning. *IEEE Internet of Things Journal, PP*(99), 1–1.
17. Lu, H., Li, Y., Chen, M., Kim, H., & Serikawa, S. (2018). Brain intelligence: Go beyond artificial intelligence. *Mobile Networks & Applications, 23*(2), 368–375.
18. Najjar, N. A. & Wilson, D. C. (2014). Differential neighborhood selection in memory-based group recommender systems. In *The International Conference of the Florida Artificial Intelligence Research Society.*
19. Ortega, F., Bobadilla, J., Hernando, A., & Gutiérrez, A. (2013). Incorporating group recommendations to recommender systems: Alternatives and performance. *Information Processing & Management, 49*(4), 895–901.
20. Quijano-Sanchez, L., Recio-Garcia, J. A., & Diaz-Agudo, B. (2010). Personality and social trust in group recommendations. In *IEEE International Conference on Tools with Artificial Intelligence* (pp. 121–126). Piscataway: IEEE.
21. Quijano-Sanchez, L., Reciogarcia, J., & Diazagudo, B. (2011). Group recommendation methods for social network environments. In *Proceedings of the Recommender System Social Web* (p. 24).
22. Quijano-Sanchez, L., Recio-Garcia, J. A., Diaz-Agudo, B., & Jimenez-Diaz, G. (2013). Social factors in group recommender systems. *ACM Transactions on Intelligent Systems & Technology, 4*(1), 8.
23. Quijano-Sánchez, L., Díaz-Agudo, B., & Recio-García, J. A. (2014). Development of a group recommender application in a social network. *Knowledge-Based Systems, 71*, 72–85.
24. Ricci, F., Rokach, L., Shapira, B., & Kantor, P. B. (2011). *Recommender systems handbook* (pp. 1–35). Berlin: Springer.
25. Sarwar, B., Karypis, G., Konstan, J., & Riedl, J. (2001). Item-based collaborative filtering recommendation algorithms. In *International Conference on World Wide Web* (pp. 285–295).

26. Schafer, J. B., Dan, F., Herlocker, J., & Sen, S. (2007). *Collaborative filtering recommender systems*. Berlin: Springer.
27. Shin, S., Jang, S. J., & Lee, S. P. (2011). The user-group based recommendation for the diverse multimedia contents in the social network environments. In *IEEE Ninth International Conference on Dependable, Autonomic and Secure Computing* (pp. 202–206).
28. Masthoff, J. (2004). Group modeling: Selecting a sequence of television items to suit a group of viewers. *User Modeling and User-Adapted Interaction, 14*(1), 37–85.
29. Xu, X., He, L., Lu, H., Gao, L., & Ji, Y. (2018). Deep adversarial metric learning for cross-modal retrieval. *World Wide Web-internet & Web Information Systems* (pp. 1–16).
30. Zhang, Y. (2016). Grorec: A group-centric intelligent recommender system integrating social, mobile and big data technologies. *IEEE Transactions on Services Computing, 9*(5), 786–795.
31. Zhang, Y., Zhang, D., Hassan, M. M., Alamri, A., & Peng, L. (2015). Cadre: Cloud-assisted drug recommendation service for online pharmacies. *Mobile Networks & Applications, 20*(3), 348–355.
32. Zhang, Y., Chen, M., Huang, D., Wu, D., & Li, Y. (2016). idoctor: Personalized and profes-sionalized medical recommendations based on hybrid matrix factorization. *Future Generation Computer Systems, 66*, 30–35.
33. Zhang, Y., Tu, Z., & Wang, Q. (2017). TempoRec: Temporal-topic based recommender for social network services. *Mobile Networks & Applications*, 1–10.

Vehicle Logo Detection Based on Modified YOLOv2

Shuo Yang, Chunjuan Bo, Junxing Zhang, and Meng Wang

1 Introduction

1.1 Research Significance

An increasing number of scholars have begun to focus on object detection in real scenes, and their results have inspired a new wave of research. The object in real scene detection has definite physical significance in road traffic problem monitoring. Extracting and detecting vehicle logos in real-world situations are important to many applications. However, these tasks remain challenging due to limited annotated data, complicated and various appearance shapes, and other problems.

1.2 Technical Difficulties

In vehicle logo detection in real scenes, the logo type and ambient condition are utilized to address the incorrect positioning of a vehicle logo. The detection problem for vehicle logos is similar to that for car plates, but diversity exists. Through a

S. Yang · J. Zhang (✉) · M. Wang
College of Electromechanical Engineering, Dalian Minzu University, Dalian, China
e-mail: zhangjunxing@dlnu.edu.cn

C. Bo
College of Information and Communication Engineering, Dalian Minzu University, Dalian, China

Key Laboratory of Intelligent Perception and Advanced Control of State Ethnic Affairs Commission, Dalian Minzu University, Dalian, China

© Springer Nature Switzerland AG 2020   75
H. Lu, L. Yujie (eds.), *2nd EAI International Conference on Robotic Sensor Networks*, EAI/Springer Innovations in Communication and Computing,
https://doi.org/10.1007/978-3-030-17763-8_8

literature [1] analysis, we summarize the difficulties and main points in vehicle logo detection.

Complex, varied, and irregular shapes can cause difficulty in feature extraction. We divided logo shapes into five types: circular (e.g., BMW and Venucia), oval (e.g., Cherry and BAIC Group), square (e.g., Honda), unusual (e.g., Suzuki), and letter pattern (e.g., Jeep).

The sizes of logos vary considerably among different brands of vehicles.

Vehicle logo locations vary. Several logos are placed in the radiator grille, and others stand on car bodies by using a spatial model.

Vehicle logo detection is influenced by light intensity effects.

The existence of noise, such as the front bumper and radiator grille, interferes with object detection for useful information. Part of a logo is connected to the radiator grille, which affects object detection.

The colors of a vehicle logo are the same for car bodies or radiator grille.

In addition, when obtaining the logo data, object detection is greatly affected by camera angle and orientation.

1.3   Typical Vehicle Logo Detection Algorithm

With regard to research on vehicle logo detection algorithms, a few classic algorithms have been developed at home and abroad. Many typical detection algorithms prioritize artificial feature extraction, after which feature models can accomplish detection tasks. For example, Psyllos and Kayafas designed a vehicle logo recognition algorithm based on the SIFT operator [2]. The SIFT feature exhibits strong robustness to image rotation, translation, illumination, and criterion. It clearly expresses the object features and executes vehicle logo detection tasks, but the computational cost is high. The SVM algorithm is used in object detection and recognition. HOG feature extraction combined with the SVM classification algorithm has achieved successful applications in pedestrian detection methods. In literature [3] and [4], the HOG + SVM algorithm was used in vehicle logo detection and recognition, and it demonstrated excellent application effectiveness. Moreover, the AdaBoost algorithm [5] has also been applied to solve multi-classification problems and has achieved good performance in vehicle logo detection that constructs weak, strong, and cascade classifiers. In artificial feature extraction, when feature expression is poor, the detection result is affected. Moreover, many manual operations can significantly increase the development time cost.

Deep-learning algorithms have developed rapidly since AlexNet [6] won the ILVRC championship in 2012. With constantly updated computer equipment, the algorithm has greatly improved in terms of speed and accuracy. After 2014, the deep-learning method was utilized in vehicle logo detection and recognition. The deep neural network has a higher precision rate of detection [7]. Meanwhile, compared with the traditional detection algorithm, the convolutional neural network (CNN) has more powerful representation capability of features [8, 9], and it also has the wide foreground of applying and developing [10–13]. Literature [14] applied a convolutional neural network to obtain object features and performed vehicle logo recognition tasks. Faster-RCNN was proposed in literature [15]. The authors reported that object detection substantially improved in speed and accuracy, and the frame rate reached 7 f/s. Tang et al. used Faster-RCNN to achieve vehicle logo detection [16]. In 2016, the YOLO method was presented in literature [17]. The YOLO method can improve the frame rate multiplier to 45 f/s. Although YOLO enhances the detection velocity, it significantly reduces the object detection accuracy. However, bringing together classification and localization tasks were thought to exert a profound effect on solving object detection problems. YOLOv2 based on the YOLO method was proposed in literature [18] in 2017, and it has answered the question of why detection veracity is poor. At 67 f/s, YOLOv2 achieves 76.8 mean average precision (MAP) on VOC 2007.

Regarding vehicle logos, this work improved the YOLOv2 network. Vehicle logo detection tasks are implemented by dimension clustering of the bounding box, reconstructing network pre-training, multi-scale detection training, and data enrichment. The limit of retrieving image data is expanded, and the contrast and noise of image data are increased and decreased, respectively, through this method to improve the accuracy and generalization of our detection algorithm.

2   Construction of Datasets

For deep learning, convolutional neural network training involves error gradient descent calculation under supervision. When training samples are sufficient, the method can obtain an optimal result. The detector of network training has low detection efficiency in small training samples. At present, public training data on vehicle logos are unavailable, and vehicle logo data are collected by researchers. This option consumes much time and energy. Through a web crawler, actual shooting, data enrichment, and network gain scheduling, this work produces datasets of vehicle logos.

2.1   Data Acquisition

By compiling web crawler scripts and practices, data on 1432 vehicles are obtained. This dataset contains 2065 vehicle logos, and 30 classes belong to common domestic car companies. The total number of images is 687 pieces, and each type of logo graphics has about 47. The category instance of images is shown in Fig. 1.

All images are in JPG form. Several pictures have a wider ratio of dissonance than the others because the actual pictures taken with the network have different

Fig. 1 Thirty classes of vehicle logo examples

sizes. To complete the image tagging work, the web crawler is scaled to a size of 600 × 400 pixels, and the actual collection image is normalized to 1400 × 1000 pixels.

2.2 Data Enrichment

The dataset is expanded in a reasonable manner because the data volume is small, and the training network model has poor generalization performance. The commonly used methods in data expansion are geometric transformation and noise increment. YOLOv2 has the function of data extension, which refers to random training adjustment for several elements, such as rotation, saturation, and hue. This adjustment can enhance the training effect on the basis of artificial data expansion. In addition, to improve the network training model for different environments for logo "sensitivity" and the generalization capability of the network model, the brightness and noise in all the training sets are adjusted. The specific adjustment is as follows.

2.2.1 Brightness Transforms

Each picture is collected by adjusting the brightness, contrast, and gamma of the image to simulate the vehicle logo detection in different light and environment conditions. Figure 2 shows the processing effect on the vehicle logo. This work uses five gamma values, namely 0.2, 0.4, 0.8, 1.2, and 2. The smaller the gamma value is, the brighter the image is. Similarly, the higher the gamma value is, the dimmer the image is. When the gamma value is 1, the image is not changed.

2.2.2 Gaussian Noise

The noise pollution in a real scene severely affects the vehicle logo detection. This work expands the vehicle logo data again by simulating the noise environment. Gaussian white noise is added to each photo, with a mean of 0 and variances of 0.06, 0.08, 0.1, 0.3, and 0.5, as shown in Fig. 3. When the variance is larger than 0.1, the noise signal can be clearly seen.

Through the processing of brightness and noise, a total of 15,752 pictures are generated in our vehicle logo data.

Fig. 2 Brightness transform of the vehicle logo object

Fig. 3 Noisy sample of the vehicle logo

3   Analysis and Improvement

3.1   YOLOv2 Algorithm

For object detection tasks, the VGG16 network is commonly used for the feature
extraction network of a deep-learning algorithm before the YOLOv2 algorithm,
but VGG16 produces redundant computation and computes complex problems.
YOLOv2 rebuilds the feature extraction network based on VGG16, and it improves
the speed of the network calculation under the premise of ensuring detection
accuracy.

3.1.1   Darknet19

Darknet19, which includes 19 convolutional layers and five max-pooling layers,
is of great importance to the YOLOv2 network. The full networks are mainly
composed of a 3×3 convolutional kernel, and the 1×1 convolutional kernel is
plugged in 3×3 convolutional layers. In this way, we can increase the depth of the
neural networks and process the object characteristics. Meanwhile, the full networks
remove dropout calculation, and YOLOv2 avoids the overfitting problem by adding
batch normalization to all the convolutional layers. Table 1 shows the comparative
performance of Darknet19 versus the VGG16 model. Overall, the velocity and
accuracy of Darknet19 are higher than those of the VGG16 model. Darknet19 has
less storage and less memory.

3.1.2   Competitive Advantage

The entire diagram information is used to extract feature maps in YOLOv2. The
system divides the input images into many grids, and the networks are calculated to
obtain the results. YOLOv2 removes Darknet19's last convolutional layers, adds
three convolutional kernels with a size of $3 \times 3 \times 1024$, increases the 1×1
kernel in each convolutional layer, and makes each feature block to have one-to-
one correspondence with the grids of the original photo. The method can hold the
spatial information of the image well without the full connected layers.

YOLOv2 uses anchor boxes from the Faster-RCNN algorithm to predict bounding boxes directly, but it runs K-means clustering on the training set bounding boxes

Table 1 Performance comparison of Darknet19 and VGG16

Model	Top-1 (%)	Top-5 (%)	CPU (s)	GPU (ms)	Weights (MB)
VGG16	70.5	90.0	4.9	10.7	528
Darknet19 (224×224)	72.9	91.2	0.66	6.4	80
Darknet19 (448×448)	76.4	93.5	2.8	11.0	80

to automatically decide the sizes and numbers. The method predicts the categories and position of the detection tasks by directly using the bounding box.

The structure of the full convolutional layers can adapt to multi-scale data training. During training, we can change the sizes of the input images with different iteration numbers. With the variety of sizes for image training, the final model detects the good effect for the object. Meanwhile, the training manner can be improved to detect small targets for YOLOv2.

3.2   Improvement Based on YOLOv2

YOLOv2 achieves good detection results on VOC 2007. However, for different target detection and environment factors, the accuracy of detection also differs, especially for our vehicle logo detection. We must improve the YOLOv2 algorithm to achieve excellent detection results for the vehicle logo. The main contents of the improvement are as follows.

3.2.1   K-Means Clustering

The anchor box is a bounding box with a different width and height. Its parameter setting exerts a tremendous effect on the accuracy of object detection. YOLOv2 runs K-means clustering on VOC 2007 to determine the statistical nature and decide the numbers of the bounding box, instead of predicting the bounding box by using handpicked priors in Faster-RCNN. Five different sizes of the bounding box are selected from the results of dimension clustering. Although the results show universality, they are not necessarily right for different data of the object detection tasks. Therefore, this work implements K-means clustering [19] again to determine the numbers of the bounding box in the data of the vehicle logo.

By counting the numbers of vehicle logos in the datasets, the range of the occurrence frequency is determined to be from 0 to 6. The cluster centers are determined randomly at 0–6, and the distance between the data stylebook and the cluster centers is calculated. When the numbers of the cluster center increase, the distance approaches a steady value. The flex point position on the curve is the best cluster number. The distance is calculated as follows:

$$d\,(\text{box}, \text{centroid}) = 1 - \text{IOU}\,(\text{box}, \text{centrid})\,. \tag{1}$$

The standard K-means results with Euclidean distance cause more errors due to the sizes of the box difference. We still use the IOU scores to calculate the distance. We select $k = 4$ as a good trade-off for the vehicle logo detection tasks, as shown in Fig. 4. We can calculate the sizes of the bounding box based on k and the sizes of the input images. When the option is 416×416 pixels, the width and height of the bounding box are (2.1696, 2.1394), (1.0520, 1.0852), (0.5455, 0.6486), and

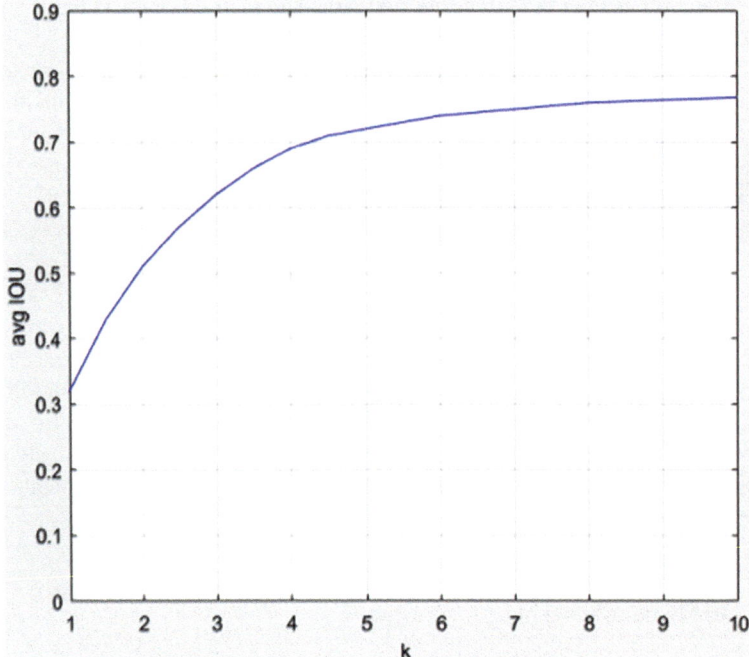

Fig. 4 Result of dimension clustering

(7.0217, 7.4663). When the option is 544 × 544 pixels, the width and height are (2.8371, 2.7977), (1.3756, 1.4190), (0.7133, 0.8481), and (9.1822, 9.7637).

3.2.2 Network Pre-Training

Many deep-learning algorithms allow neural networks to be pre-trained on ImageNet or VOC. Then, the pre-training model is used for training again in solving practical problems. Network pre-training improves the feature extraction capability and speed of the network to help with object detection. However, the classification of ImageNet varies, and it is not suited for vehicle logo pre-training. Thus, we change the process of pre-training as follows.

ImageNet data are used for pre-training in Darknet19.

Using the 600 × 400 data of vehicle logo to train Darknet19 enables the networks to adapt to the feature of vehicle logo.

The 1400 × 1000 data of vehicle logo are used to train Darknet19. The point is to let the network adapt to the feature of vehicle logo and multi-scale detection tasks.

Through network pre-training, the network becomes familiar with the feature of vehicle logo. Thus, the method can help achieve detection tasks.

3.2.3 Multi-Scale Detection Training

The center dimension can be changed flexibly because Darknet19 only has convolutional and pooling layers. By using multi-scale training, we can increase the robustness of the network, the accuracy of the detector, and the generalization of the model. Darknet19's convolutional layers downsample the image by a factor of 32. 416×416 is the minimum dimension of our network input, namely the size of the feature map is 13×13. Thus, we arrive at the size of network formula as follows:

$$S_{\text{area}} = (13 + i) \times 32, \tag{2}$$

where i represents random natural numbers between 0 and 6. We select seven large sizes of the network input because the size of data is very large. The fixed inputs of the network have low accuracy for small targets, and multi-scale training can improve the detection accuracy for vehicle logos of different sizes.

4 Experimental Design and Interpretation of Results

4.1 Comparative Experiment

To enhance the detection effect of our algorithm, we use three algorithms in contrastive experiments. These three algorithms are based on the AdaBoost cascade classifier, the HOG + SVM algorithm, and Faster-RCNN. The comparison between the improved algorithm and the original algorithm of YOLOv2 shows that the improvement in this work is effective in practical evaluation.

AdaBoost and SVM are classic machine learning algorithms that work well in small sample target detection and recognition. They have been applied in vehicle detection. This work compares the traditional learning method with the deep-learning method in terms of detection results. Faster-RCNN has been used many times in small target detection and recognition. Meanwhile, the network can use GPU to accelerate the operation, so it is used as the main contrast object.

4.2 Experimental Results and Analysis

4.2.1 Comparison Experiment on the Four Algorithms

The final test result is the standard in this work because of the difference between the algorithm and training method. The experiment is conducted on 1432 and 15,752 datasets of the algorithm and training method, respectively. The ratio of the training and testing samples is 8:2. The standard in this study is the average overlap rate and time consumed for each photo. Table 2 shows the results of different algorithms.

Table 2 The comparison of the four algorithms

Algorithm	Average overlap rate		Time consuming/s
	Original data	Extensible data	
AdaBoost	0.793	0.612	0.33
HOG + SVM	0.810	0.744	0.24
Faster-RCNN (VGG16)	0.854	0.852	1.5
YOLOv2	0.823	0.825	0.03

Table 2 shows that the deep-learning algorithm has high stability in the detection tasks of the vehicle logo and can overcome the effect of noise. AdaBoost and SVM rely mainly on the extraction of data features in the early stage but easily generate detection errors for large noise samples. Faster-RCNN spends much time in data analysis, uses a large size of the network input, and increases detection accuracy to reduce the processing speed. Meanwhile, YOLOv2 is more than ten times faster than the other algorithms on the basis of maintaining accuracy, and it is the best choice in the vehicle logo detection tasks.

4.2.2 Comparison Experiment on the Improved Algorithm

In this study, three aspects of the algorithm, namely dimension clustering, network pre-training, and multi-scale training, are improved. The results of the comparison of the performance of Faster-RCNN, YOLOv2, and improved algorithms are shown in Table 3.

The contrastive analysis shows that we can achieve good results by using few candidate boxes after resetting the dimension clustering. The high detection overlap rate is maintained on the basis of reducing the amount of calculation and resource occupancy. Using the multi-vehicle-logo pre-training model, the algorithm can achieve good results in object detection, and the capability of the network to extract the vehicle logo features is improved. We are unable to perform multi-scale training because Faster-RCNN has a full connected layer. Through training iteration to change the input network scale, the network of detection acquires enhanced adaptability for multi-scale logo pictures, and it exhibits improved detection accuracy for the vehicle logo in the real scene. With the increasing size of the network input, the network greatly improves the MAP value of the vehicle logo. The experiments prove the apparent effect of the improved algorithm.

Acknowledgments This work is supported by National Key Technology Research and Development Program of the Ministry of Science and Technology of China (No. 2015BAD29B01), Key Research Guidance Plan Project of Liaoning Province (No. 2017104013), Natural Science Foundation of Liaoning Province (No. 201700133), and Fundamental Research Funds of Central University (No. 0102-20000101).

Table 3 The comparison of improved algorithm

Algorithms	Number of anchors	Average overlap rate	Mean average precision (30 classes)					
			Single-scale (544 × 544)			Multi-scale (416–608)		
			Original model	Improved model	Time/s	Original model	Improved model	Time/d
Faster-RCNN	9	0.852	0.723	–	0.35	–	–	–
YOLOv2	5	0.825	0.747	0.762	0.02	0.755	0.771	0.03
The modified YOLOv2	4	0.834	0.748	0.744	0.02	0.764	0.785	0.03

References

1. Du, S., Ibrahim, M., Shehata, M., et al. (2013). Automatic license plate recognition (ALPR): A state-of-the-art review. *IEEE Transactions on Circuits and Systems for Video Technology, 23*(2), 311–325.
2. Psyllos, A. P., & Kayafas, E. (2010). Vehicle logo recognition using a SIFT – Based enhanced matching scheme. *IEEE Transactions on Intelligent Transportation Systems, 11*, 322–328.
3. Llorca, D. F., Arroy, O. R., & Sotelo, M. A. (2013). Vehicle logo recognition in traffic images using hog features and SVM. In *IEEE Conference on Intelligent Transportation Systems* (pp. 2229–2234).
4. Sun, Q., Lu, X., Chen, L., et al (2014). An improved vehicle logo recognition method for road surveillance images. In *IEEE Proceedings of the 2014 Seventh International Symposium on Computational Intelligence and Design* (pp. 373–376).
5. Sam, K. T., & Tian, X. L. (2012). Vehicle logo recognition using modest AdaBoost and radial Tchebichef moments. In *International Conference on Machine Learning and Computing* (pp. 91–95).
6. Krizhevsky, A., Sutskever, I., & Hinton, G. E. (2012). ImageNet classification with deep convolutional neural networks. *International Conference on Neural Information Processing Systems, 60*(2), 1097–1105.
7. H, L., Li, Y., Chen, M., Kim, H., & Serikawa, S. (2018). Brain intelligence: Go beyond artificial intelligence. *Mobile Networks and Application, 23*(2), 368–375.
8. Serikawa, S., & Lu, H. (2014). Underwater image dehazing using joint trilateral filter. *Computers and Electrical Engineering, 40*(1), 41–50.
9. Lu, H., Li, Y., Uemura, T., Kim, H., & Serikawa, S. (2018). Low illumination underwater light field images reconstruction using deep convolutional neural networks. *Future Generation Computer Systems, 82*, 142–148.
10. H, L., Li, Y., S, M., Wang, D., Kim, H., & Serikawa, S. (2018). Motor anomaly detection for unmanned aerial vehicles using reinforcement learning. *IEEE Internet of Things Journal, 5*(4), 2315–2322.
11. H, L., Li, B., Zhu, J., Li, Y., et al. (2017). Wound intensity correction and segmentation with convolutional neural networks. *Concurrency and Computation: Practice and Experience, 29*(6), e3927.
12. X, X., He, L., H, L., Gao, L., & Ji, Y. (2019). Deep adversarial metric learning for cross-modal retrieval. *World Wide Web, 22*(2), 657–672. https://doi.org/10.1007/s11280-018-0541-x.
13. Li, P., Wang, D., Wang, L., & Lu, H. (2017). Deep visual tracking: Review and experimental comparison. *Pattern Recognition, 76*, 323–338.
14. Huang, Y., Wu, R., Sun, Y., Wang, W., & Ding, X. (2015). Vehicle logo recognition system based on convolutional neural networks with a pretraining strategy. *IEEE Transactions on Intelligent Transportation Systems, 16*(4), 1951–1960.
15. Ren, S., He, K., Girshick, R., & Sun, J. (2017). Faster R-CNN: Towards real-time object detection with region proposal networks. *IEEE Transactions on Pattern Analysis & Machine Intelligence, 39*(6), 1137.
16. Tang, T., Zhou, S., Deng, Z., et al. (2017). Vehicle detection in aerial images based on region convolutional neural networks and hard negative example mining. *Sensors, 17*(2), 336.
17. Redmon, J., Divvala, S., Girshick, R., & Farhadi, A. (2016). You only look once: Unified, real-time object detection. In *Proceedings of the IEEE Conference on Computer Vision and Pattern Recognition* (pp. 779–788).
18. Redmon, J., & Farhadi, A. (2017). YOLO9000: Better, faster, stronger. In *Proceedings of the IEEE Conference on Computer Vision and Pattern Recognition* (pp. 7263–7271).
19. Huang, Z. (1998). Extensions to the k-means algorithm for clustering large data sets with categorical values. *Data Mining and Knowledge Discovery, 2*(3), 283–304.

Energy-Efficient Virtual Machines Dynamic Integration for Robotics

Haoyu Wen, Sheng Zhou, Zie Wang, Ranran Wang, and Jianmin Lu

1 Introduction

With the rapid development of the Internet, the demand for computational capabilities of various scientific data and commercial data far exceeds the computing capacity of its own data center [1]. It is precisely because of this demand that it has promoted the rapid development of cloud computing technology. Cloud computing is a dynamic and scalable computing method. It uses virtualized computer resources as a service and provides it to users through the network [2]. This model can help computers allocate resources on demand, usually with a dynamic expansion and distributed features. These cloud data centers that are dependent on the cloud computing environment are also bringing about significant power consumption and CO_2 emissions. It is estimated that from 2005 to 2010, the power consumption of data centers in the world has increased by 56%, accounting for 1.1–1.5% [3] of global electricity consumption in 2010. Moreover, unless the current conventional resource management scheme is changed to achieve efficient energy use, data center energy consumption will continue to grow rapidly. This method can also be used in drone anomaly detection [4]. Similarly, in the artificial intelligence, social network services, mobile healthcare, Internet of things, and other fields [5–9], the amount of data is huge, resulting in a lot of energy consumption.

The current problem faced by cloud data centers is that the resource usage of CPUs and other resources is usually only 10–50% of the total resources. Such a reserved configuration method results in the waste of a large amount of extra power resources and even causes many idle servers to continue to be consumed [2]. During

H. Wen (✉), S. Zhou, Z. Wang, R. Wang, J. Lu
School of Information and Safety Engineering, Zhongnan University of Economics and Law, Wuhan, China
e-mail: haoyuwen.zuel@qq.com

© Springer Nature Switzerland AG 2020

87

H. Lu, L. Yujie (eds.), *2nd EAI International Conference on Robotic Sensor Networks*, EAI/Springer Innovations in Communication and Computing,
https://doi.org/10.1007/978-3-030-17763-8_9

the operation of a cloud data center, a virtual machine is allocated and scheduled as a computing resource having a cloud computing platform. The virtual machine provides a user with a computing resource, and the virtual machine itself needs to run on a physical host. According to the Open Compute Project (OCP) report, the power usage effectiveness of the Facebook data center in Prairie in the fourth quarter of 2015 reached 1.09, while in the Forest City PUE reached 1.08. This means that computing resources consume about 91% of all consumed resources in the cloud data center. Therefore, to solve the power consumption problem, the focus is on improving the utilization of computing resources of each node in the cloud data center. This is similar to the study by Serikawa et al. [10]. They proposed a new joint triangular filter with fast and non-approximate constant time algorithms [11]. In addition, in order to speed up the operation of the algorithm, for example, the most advanced level set method can well segment the object [12]. However, this method is time consuming and inefficient, so that balancing the resource load can solve the problem well. Then, our method can also be applied to deep learning image detection [13].

The main work of this paper will focus on the study of virtual machine integration systems. Taking OpenStack as an example, the integration of virtual machines mainly involves the scheduling of control nodes and the assignment and migration of virtual machines on compute nodes. The computing nodes need to make decisions on virtual machines. The necessity of migration and the time of migration to avoid the "jitter" caused by excessive virtual machine migration result in a decline in the overall performance of the cloud data center. The main purpose of the integration of virtual machines is to save energy consumption as much as possible while achieving cloud data center load balancing. For this purpose, this paper studies the forecasting effect of the Markov chain process model on time series. Based on Pearson correlation coefficient, a K-sequence mixed Markov model is proposed to predict the CPU usage of the host. Based on the CloudSim [14] cloud computing platform simulation Java package released by the University of Melbourne, Australia, a simulation experiment based on hybrid Markov-based host load forecasting is implemented. In the simulation experiment, the number of virtual machine migrations, data center energy consumption, SLA violations, etc. are compared with traditional threshold-based average load detection algorithms [2] and local regression robust (LRR) detection algorithms [15]. Through simulation experiments, it is verified that the host load forecast model algorithm proposed in this paper can effectively reduce the number of virtual machine migrations and energy consumption of cloud data centers.

The content of this article is organized as follows: The second part introduces the related work and research status. The third part proposes the model and explains the implementation of the algorithm. The fourth part and the fifth part introduce the design of the experiment and the result analysis. The sixth part summarizes the full text.

2 Related Works

The main purpose of dynamic integration is to use the real-time migration method to reallocate virtual machines to fewer data center nodes by considering real-time requests from virtual machines for resources [16] and to switch idle node hosts to low-power states. This will improve the use of physical resources and reduce energy consumption. The distribution and integration of virtual machines in the cloud platform have two modes, namely static mode and dynamic mode. In the static integration of virtual machines, scheduling and integration algorithms are often designed based on average historical resource utilization and user-defined performance indicators [17, 18]. However, this method assumes that the virtual machine resource requirements are known in advance and does not take into account changes in the virtual machine workload. Nathuji et al. [19] discussed the energy benefits brought by the integration of dynamic virtual machines and found that overall energy consumption can be significantly reduced. Beloglazov et al. [20] divided the issue of dynamic virtual machine integration into the following four sub-questions after extensive research on virtual machine integration technologies:

(1) Determine when the node enters a low-load state so that all virtual machines can be migrated from the node. The node can also be switched to a low-power mode, such as sleep mode.
(2) Determine when the node enters a high-load state. This should select the appropriate virtual machine and migrate to other suitable active nodes to avoid server performance degradation.
(3) Select a suitable virtual machine from a high-load host for migration.
(4) Find a suitable placement place on the host of other activities and migrate the virtual machine to be migrated to this host.

The current research on the decision-making of physical host high/low-load states can be roughly divided into the following three methods:

The first is to make decisions using a threshold-based heuristic.

ATWood et al. [21] studied the dynamic migration of virtual machines. They used a threshold-based method to determine that a physical host in the data center entered a high-load state and then migrated the virtual machine out of it to meet the load-balancing requirements purpose. Although this kind of threshold-based load detection algorithm is relatively simple, once the physical host's workload suddenly changes, for example, the host's load status frequently enters high-load status and low-load status within a short period of time, the host may be busy. Migration of virtual machines, and excessive migrations can cause the host to become busy and degrade its performance, which is a serious violation of the service level agreement (SLA). Zhu et al. [22] studied a large number of dynamic virtual machine integration problems. They used a static threshold heuristic method to set the CPU usage threshold to 85%. Once the node CPU usage exceeded the threshold, the node was determined to enter a high-load status [2]. Gmach et al. [23, 24] also did a similar study. They analyzed the change track of workloads in cloud data centers and set the CPU usage threshold to 85%. However, this method of static threshold heuristic setting does not apply to hosts whose workload changes dynamically or where unknown random changes occur.

The second is to periodically adapt the virtual machine placement without load detection.

Verma et al. [25] simulated the dynamic virtual machine consolidation problem as a boxing problem. They considered the consumption of virtual machine migration and proposed a heuristic method to minimize data center power consumption. However, they only periodically adjust the virtual machine position without using any algorithm to determine the optimal placement of the virtual machine. If the host placed by the virtual machine enters a high-load state at the next moment, the virtual machine may need to be migrated again. This leads to frequent migration of virtual machines. Weng et al. [26] proposed a load-balancing system for periodically reducing the consumption of virtual machine allocations in a cluster to detect high-load and low-load physical hosts and redistribute virtual machines. However, in large data centers, the number of physical hosts and virtual machines is very large. This system will inevitably cause a lot of extra energy consumption when redistributing virtual machines.

The third type is based on statistical analysis of the resource usage of historical CPUs and so on to make predictions on the next moment.

Bobroff et al. [27] proposed a server load forecasting algorithm based on the time-series analysis of host historical data. However, the algorithm is too complicated and its time complexity and space complexity are too high. Huang et al. [28] comprehensively considered the dynamic fluctuations and resource conflicts of each virtual machine workload on the cloud computing system. Based on the historical workload data of the virtual machine, the autoregressive integrated moving average (ARIMA) model was used. Future virtual machine resource requests are predicted. ARIMA is a time-series predictive model that can predict the resource utilization of physical host nodes at the next point in time through historical data. However, in the study of virtual machine integration, it is often only necessary to determine the state of the CPU and accurately predict its usage. Often errors are large and can also result in higher runtimes. Beloglazov et al. [20] did a lot of research on the forecasting work of the Markov chain model on the host's future workload, and used the Markov model to predict whether the host machine will enter a high-load state in the future. However, the author of this paper only considers the influence of CPU state on the next moment at the current moment, and does not consider the CPU state before the current moment will also affect the next moment, so the ordinary single-order Markov chain model is in the future for some time. The host load forecast can be very error. This article will study prediction methods based on statistical analysis.

3 *K*-Order Mixed Markov Model

The Markov model is a widely used predictive model, which is mainly based on historical discrete data to predict the future moment. Assuming a random process, the parameter set T of the random process is a discrete-time set $\{X_n, n \in T\}$, that is, $T = \{0, 1, 2, \ldots\}$, and the state space of the entire possible composition of the random process is a discrete state set $I = \{i_0, i_1, i_2, \ldots\}$.

Definition 1 If the random process $\{X_n, n \in T\}$ for any non-negative integer and arbitrary $i_0, i_1, \ldots, i_{n+1} \in I$, the conditions satisfy

$$\{X_n, n \in T\} \tag{3.1}$$

We call $\{X_n, n \in T\}$ the Markov chain.

Definition 2 $\forall i, j \in S$, then call $P\{X_{n+1} = j | X_n = i\} = p_{ij}(n)$ as one-step transition probability at time.

If $\forall i, j \in S$, $p_{ij}(n) \equiv p_{ij}$, that is, p_{ij} is independent of n, it is said that the transition probability is stable. At this time, $\{X_n, n \in T\}$ is called a time-homogeneous Markov chain, and $P = (p_{ij})$ is a one-step transition probability matrix

$$P = \begin{pmatrix} p_{00} & p_{01} & \cdots & p_{0j} & \cdots \\ p_{10} & p_{11} & \cdots & p_{1j} & \cdots \\ \vdots & \vdots & \ddots & \vdots & \vdots \\ p_{i0} & p_{i1} & \cdots & p_{ij} & \cdots \\ \vdots & \vdots & \cdots & \vdots & \ddots \end{pmatrix}, p_{ij} \geq 0 (\forall i, j \geq 0), \sum_{i=0}^{\infty} \sum_{j=0}^{\infty} p_{ij} = 1 \tag{3.2}$$

For the host resource usage data collected at regular intervals, this is a discrete-time Markov chain. Actually, the state transition probability is not related to the current moment, so it can be described as a time-homogeneous discretization time Markov chain [29].

Since the accuracy of ordinary single-order Markov model prediction is often not high, and it is difficult to meet the requirements, the literature [30] combines the correlation rules and proposes a new hybrid Markov model, which greatly improves the accuracy of prediction. Mukund et al. [31] comprehensively considered the *K*th-order Markov model and proposed a Markov model on the basis of this, and proved that the model can accurately predict and have a higher coverage, but with the increase of *K* value. The complexity of the algorithm will increase dramatically. Considering the persistence of computer programs, the current use of computer resources is affected by the running process of the program. So the current state may not only be related to the previous state, but related to the previous state. The ordinary Markov model considers only one state before the current state, and if all the states before the current state are considered, the time and space complexity of the problem calculation is greatly increased, so only K before the current time can be considered. State ($K < n$) and hypothetical:

$$P\{X_{n+1} = i_{n+1}|X_0 = i_0, X_1 = i_1, \ldots, X_n = i_n\}$$
$$= P\{X_{n+1} = i_{n+1}|X_n = i_n, X_{n-1} = i_{n-1}, \ldots, X_{n-(k-1)} = i_{n-(k-1)}\} \quad (3.3)$$

Conforming to (3.3) is the K-order Markov model. When K is 1, it degenerates into an ordinary single-order Markov model. The state space of the K-order Markov model is

$$S_i^K = \{i_{l-(K-1)}, i_{l-(K-2)}, \ldots, i_l\} \quad (3.4)$$

The K-order Markov model has three evaluation parameters [31], namely the accuracy rate, the number of states, and the coverage rate. Accuracy is used to assess the accuracy of the prediction process; the state quantity refers to the size of the state space of the K-order Markov model. Since the model state space is closely related to the time and space complexity of the model algorithm, this parameter is also used to evaluate the execution time of the model algorithm; coverage refers to the fact that when the value of K is large, the state space increases rapidly and the historical data is limited, so many states in the state space may never occur, i.e., they are limited. The historical data set cannot cover all the states in the state space, and the coverage ratio is equal to the ratio between the number of historical data and the number of state spaces.

The literature [31] has proved that if the K value of the model is increased, the accuracy rate can be effectively improved, but at the same time, because of the increase of the K value, the state space also reaches exponential growth, and the model coverage rate also decreases sharply. This will also affect the accuracy of the forecast.

Weighted Markov models are widely used in the various fields of prediction algorithms [32, 33]. For the time-homogeneous discrete-time Markov model, the effect of the non-synchronous long (latency) on the next moment is different. The weight of this influence is calculated using the Pearson correlation coefficient.

For the K-order Markov model, the states of the K moments before the current moment have different effects on the current moment. The Pearson correlation coefficient, also known as the linear correlation coefficient, can be used to characterize the interaction weights of states at different moments. Its formula is

$$r = \frac{\sum\limits_{i=1}^{n} (X_i - \bar{X})(Y_i - \bar{Y})}{\sqrt{\sum\limits_{i=1}^{n} (X_i - \bar{X})^2 \times \sum\limits_{i=1}^{n} (Y_i - \bar{Y})^2}} \quad (3.5)$$

Equation (3.5) is used to calculate the correlation coefficient between variables X and Y. When used to calculate the correlation coefficient at different times of the same sequence, it can be converted into an autocorrelation coefficient, that is, replace the $Y_i - \bar{Y}$ in 2.5 molecules with $X_{i+z} - \bar{X}$, and the formula is as shown in (3.6):

$$r_z = \frac{\sum_{i=1}^{n-z} \left(X_i - \bar{X}\right)\left(X_{i+z} - \bar{X}\right)}{\sum_{i=1}^{n} \left(X_i - \bar{X}\right)^2} \tag{3.6}$$

which represents the zth order autocorrelation coefficient (latency is the correlation between z times and the current time).

The Pearson autocorrelation coefficient characterizes the correlation of the same variable with each other at different times. This value may be an integer or a negative number or 0. What is needed in practice is the influence of the variable's non-synchronized value on the current value, so it needs to be normalized to obtain various time-delay weights.

$$w_z = \frac{|r_z|}{\sum_{z=1}^{j} |r_z|} \tag{3.7}$$

where j represents the maximum order to be calculated.

4   Host Model

4.1   Model Establishment

During the running of the computer, a part of resources (such as CPU, memory, etc.) is used at each moment, thereby generating the workload of the computer. In the cloud platform, when each virtual machine on the physical host is running, only a part of CPUs, RAMs, and other resources of the physical host is used. The workload of the physical host is composed of the consumption of CPUs and other resources generated after the creation of a group of virtual machines. At the beginning, we assume that the CPU usage of the host measured at a series of discrete-time points can be described as a time-homogeneous discrete-time Markov chain (DTMC).

Assuming that $U = \{u_1, u_2, u_3, \ldots, u_n\}$ is the observed historical CPU usage data sequence, the observation interval time is t, n is the total number of current data, the CPU status is $C = \{c_1, c_2, \ldots, c_m\}$, m is the number of states divided by the CPU, and C_l^i is used to represent the CPU time at t_l. The state is c_i. Let the current moment be S^k, and the time to be predicted is t_{l-1}. The state space of the K-order Markov model can be described as:

$$S = \left\{ \left(S^k, c_1 \right), \left(S^k, c_2 \right), \ldots, \left(S^k, c_m \right) \right\} \tag{4.1}$$

among them, $S^k = \left(C_{l-k}^x, C_{l-k-1}^y, \ldots, C_{l-1}^z \right), 1 \le x, y, z \le m$.

It can be seen from Eq. (4.1) that the number of states in the current state space is m^k. Because the CPU state change sequence can be arbitrarily long, for CPU data, the number of historical CPU data sequences used to calculate the conditional probability cannot be infinite, and as the sequence grows, the number of states of the Markov model grows exponentially, which leads to extremely high runtimes. Therefore, $K << n$ is often chosen as the value, and the length of the CPU data sequence is n.

Through the defined state space and historical CPU data, according to the maximum likelihood rule, the Markov transition probability can be calculated as

$$P\left(c_i | S^k \right) = \frac{\text{Frequency} \left(\left(S^k, c_i \right) \right)}{\text{Frequency} \left(S^k \right)} \tag{4.2}$$

where Frequency $\left(S^k \right)$ denotes the number of occurrences of the sequence S^k, and Frequency $\left(\left(S^k, c_i \right) \right)$ denotes the number of occurrences of the state c_i immediately following the S^k. The calculated transition probability matrix P is the m^k order. m^k refers to the model when the number of CPU states is m. The number of state spaces can be up to m^k.

For the K-order Markov model, assume that the state sequence $\left(C_{l-k}^1, C_{l-k-1}^1, \ldots, C_{l-1}^1 \right)$ is called state S_0 and the state sequence $\left(C_{l-k}^1, C_{l-k-1}^1, \ldots, C_{l-1}^2 \right)$ is called the state S_1, and so on, the sequence of states $\left(C_{l-k}^m, C_{l-k-1}^m, \ldots, C_{l-1}^m \right)$ is called the state S_{m^k-1}. First consider P_{00}, the probability of transition from state S_0 to state S_0:

$$P_{00} = P \{ S_0 | S_0 \}$$

$$= P \left\{ \left(C_{l-k-1}^1, C_{l-k-2}^1, \ldots, C_l^1 \right) | \left(C_{l-k}^1, C_{l-k-1}^1, \ldots, C_{l-1}^1 \right) \right\} \tag{4.3}$$

For the state sequence, since the first k-1 states in $\left(C_{l-k-1}^1, C_{l-k-2}^1, \ldots, C_l^1 \right)$ have already been determined to occur, there will actually be $P_{00} = P\{S_0|S_0\} = P\{C_l^1 | \left(C_{l-k}^1, C_{l-k-1}^1, \ldots, C_{l-1}^1 \right)\}$, and the transition probability is 0 for the determination of the non-occurring state sequence. So the following formula can be drawn:

$$P_{ij} = \begin{cases} 0, & \text{If the first } k-1 \text{ and the last } k-1 \text{ states are different} \\ P\left\{C_l^i | S_j\right\}, & \text{If the first } k-1 \text{ and the last } k-1 \text{ are in the same state} \end{cases}$$
(4.4)

Then the formula can calculate the transition probability matrix P, where P is an m^k matrix

$$P = \begin{pmatrix} P\left(c_1 | S_0\right) \cdots & 0 & \cdots P\left(c_j | S_0\right) & \cdots & P\left(c_m | S_0\right) \\ \vdots & \ddots & \vdots & & \vdots \\ 0 & \cdots P\left(c_i | S_{m^k-1}\right) & \cdots P\left(c_j | S_{m^k-1}\right) & \cdots & P\left(c_m | S_{m^k-1}\right) \end{pmatrix}$$
(4.5)

The Markov multi-step transition probability matrix $P^{(n)}$ can be further calculated as

$$P^{(n)} = P^n$$
(4.6)

For the K-order Markov model, assuming that only the load state of the next n moments is predicted, the transition probability matrix needs to be calculated to n steps, i.e., P^n. Considering that the K-order Markov transition matrix describes the transition probability of the state of K moments at this moment in time, we need to calculate the weight of the current state of the n state space before the current moment, and the step length reaches $K+n-1$. That is, the $K+n-1$ order Pearson autocorrelation coefficient is calculated.

Assume that the $K + n + 1$ order Pearson correlation coefficients are $r_1, r_2, r_3, \ldots, r_{K+n-1}$, respectively, and then calculate the weight of each step sequence. For a resource data sequence $U_i = \{u_{i-K+1}, \ldots, u_{i-1}, u_i\}$ with a step size i, the corresponding Pearson correlation coefficient set is $\{r_{i-K+1}, \ldots, r_{i-1}, r_i\}$, so the order weights are obtained as

$$W_i = \frac{|r_{i+K-1}| + |r_{i+K-2}| + \ldots + |r_i|}{\sum\limits_{j=1}^{n+K-1} |r_j|}$$
(4.7)

After the weights are standardized, the weights of each step are

$$w_z = \frac{W_z}{\sum\limits_{j=1}^{n} W_j}, 1 \leq z \leq n$$
(4.8)

Assume that the transition from the sequence h in step i to the state c_j is represented by $P_{hc_j}^i$, and the probability of predicting the state c_j at the next time point is

$$P_{c_j} = \sum_{i=1}^{n} w_i P_{hc_j}^i \qquad (4.9)$$

The state at the next moment c_j means that the process is in the state $S^* = \langle S^k, c_j \rangle$, and $S^k = (C_{(l-k)}^x, C_{(l-k-1)}^y, \ldots, C_{(l-1)}^z), 1 \leq x, y, z \leq m$ has been specified above. So $P_{hc_j}^i$ can be calculated by Eq. (4.10):

$$P_{hc_j}^i = \sum P_{S^k c_j}^i \qquad (4.10)$$

where $P_{S^k c_j}^i$ represents the probability of transfer from S^k to c_j in the i-step transition matrix.

The final prediction result is the probability of each state of the CPU at the next moment. This situation is often encountered: two or more states have higher probability of occurrence, and the probability difference is smaller. If blindly selecting the state with the highest probability of occurrence as the CPU state at the next time point, the accuracy is not necessarily high. Consider setting a threshold. When the probability difference between the state with the highest probability and the state with the smallest probability exceeds this threshold, the CPU will be considered to enter this state at the next moment.

4.2 Algorithm Design

The state space of the K-order Markov model is the K-time combination of the CPU state space, and the size is the K power of the size of the CPU state space. After determining the K-order Markov state space, the probability of each Markov state transition to other Markov states (including the transition to itself) is calculated according to the historical CPU state. After the combination, the one-step transition probability matrix is obtained. Then calculate the similarity between the CPU states of each step, that is, the Pearson correlation coefficient, and then determine the impact weight of the CPU state of each step on each state that the CPU may be in the next moment. When K is greater than 1, the state of the CPU at the next moment is influenced by the previous K states. Therefore, it is necessary to correct the previously calculated weights of the joint weights of the K states. Through the weights and transition probability matrix, the probability of each state of the CPU at the next moment can be predicted. Similarly, the CPU status at the next time (multiple times) can be predicted. The forecast value for each moment must be tested.

Algorithm 1 K-stage mixed Markov model based prediction algorithm

Input: cpuUtilizationHistory, cpuStat, K, n, overloadStat
Output: Whether the host may be overloaded during the next period of time
1: $cpuStatSeq \leftarrow getStatClassification(cpuUtilizationHistory, cpuStat)$
2: $statSpace \leftarrow generateStatSpace(cpuStat)$
3: $statSeq \leftarrow generateStatSeq(cpuStatSeq, K)$
4: $double[][]p \leftarrow newdouble[statSpace.length][statSpace.length]$
5: $int[]count \leftarrow newint[spaceLen];$
6: **for** $i \leftarrow 0$ to $statSpace.length$: **do**
7:    $total \leftarrow 0$
8:    **for** $j \leftarrow 0$ to $statSpace.length$: **do**
9:       $flag \leftarrow 0$
10:      **for** $k \leftarrow 0toK$: **do**
11:         **if** $(seq[j][k]! = statSpace[i][k]$ **then**
12:            $flag \leftarrow 1$
13:            $break;$
14:         **end if**
15:         **if** $flag = 0$: **then**
16:            $++total$
17:         **end if**
18:      **end for**
19:   **end for**
20:   **for** int $j \leftarrow 0$ to $statSeq.kength - K$: **do**
21:      $flag \leftarrow 0$
22:      **for** $k \leftarrow 0toK$: **do**
23:         **if** $seq[j][k]! = statSpace[i][k]$ **then**
24:            $flag \leftarrow 1$
25:            $break;$
26:         **end if**
27:      **end for**
28:      **if** $flag = 0$: **then**
29:         $index \leftarrow statSpace.indexOf(statSeq[j + 1]);$
30:         $++count[index];$
31:      **end if**
32:   **end for**
33:   **for** int $j \leftarrow 0$ to $statSpace.length$: **do**
34:      **if** $total = 0$ **then**
35:         $p[i][j] \leftarrow 0;$
36:      **else**
37:         $p[i][j] \leftarrow (double)count[j]/total$
38:      **end if**
39:   **end for**
40: **end for**
41: **for** $i \leftarrow 1$ to n **do**
42:    $p[i] \leftarrow Matrix.pow(p, i)$
43: **end for**
44: $Calculate\ the\ probability\ matrix\ transferred\ to\ each\ CPU\ state\ in\ the$
45: $ith\ transition\ probability\ matrixStore\ it\ in\ sumProbability[i]$
46: $r \leftarrow calculateCorrelation(cpuUtilizationHistory, z, K)$
47: $w \leftarrow calculateWeight(r)$
48: **if** $K > 1$ **then**
49:    $Correction\ weight\ w$
50: **end if**

```
51:  stepProbability ← sumProbability * w
52:  for i in stepProbability : do
53:     m = max(i)
54:     if max(i) = overloadedStat : then
55:        s = second_largest(i)
56:        confidence_threshold = mC1.96 * square((m * (1 − m))/p)
57:        if m − s > confidence_threshold then
58:           return true
59:        end if
60:     end if
61:  end for
62:  return false
```

5 Experiment and Result Analysis

5.1 The Experimental Process

Simulation Tools and Experiment Environment CloudSim is actually a Java package that can be used to simulate resource management and scheduling experiments in a cloud computing environment. CloudSim simulates the distribution of virtual machines (including CPU, memory, storage space, and bandwidth) on the underlying analog cloud data center and provides interfaces at the upper layer. Users can use code emulation to create cloud data centers and invoke interfaces to create virtual machines.

This experiment program is written in Java language. The operating environment is shown in Table 1.

Simulation Process In order to ensure the reliability of the experimental results, this experiment will use the CPU utilization data from the actual system instead of the randomly generated data. Park [34] and others introduced the project CoMon, which is a project to collect CPU utilization data of thousands of deployed virtual machines from hundreds of servers in the world. Unfortunately, for various reasons the project has completely failed. This experiment uses some of the data collected before this project. The data collected by the CoMon project is an interval of 5 min. One day is a group of 288 data. Figure 1 shows the CPU data for four random VMs.

The simulation parameters used in the experiment are shown in Table 2.

Table 1 Experimental environment configuration

Configuration name	Details
CPU	Intel(R) Xeon(R) CPU E5–2620 v2@2.10 GHz
RAM	64 GB
Operating system	64bit CentOS 7
JDK version	OpenJDK 64–Bit Server VM (build 1.8.0_65–b17)

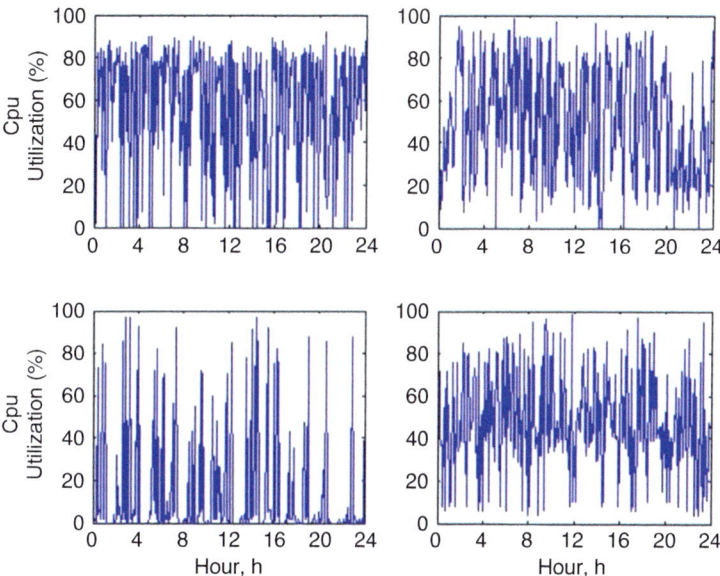

Fig. 1 CPU 24 h utilization (in 5 min intervals)

Table 2 Experimental parameter settings

	Parameter	Value
Host	CPU	Intel(R) Xeon(R) CPU E5–2620v22.1 GHZ
		Intel(R) Xeon(R) CPU E5–24201.9 GHZ
	RAM	32 GB RAM
		16 GB RAM
	Quantity	800
Virtual machine	CPU	2000 MIPS, 1600 MIPS, 1200 MIPS, 800 MIPS
	RAM	2048 MB, 3072 MB, 4096 MB, 1024 MB
	Quantity	3596

At the beginning of the execution of the prediction process, the CPU state interval is first defined. We divide the CPU usage rate from 0% to 100% into 10 states, and define the state 1 and the state 2 as the low-load states, and the state 9 and the state 10 as high-load states.

The CPU status is divided as shown in Table 3.

According to Table 3 CPU status can determine the status of the CPU at each moment.

Table 3 CPU status interval

Status	CPU usage rate	Upper CPU usage
1	[0%, 10%)	10%
2	[10%, 20%)	20%
3	[20%, 30%)	30%
4	[30%, 40%)	40%
5	[40%, 50%)	50%
6	[50%, 60%)	60%
7	[60%, 70%)	70%
8	[70%, 80%)	80%
9	[80%, 90%)	90%
10	[90%, 100%)	100%

5.2   Analysis of Simulation Results

In the simulation experiment, we tested the traditional threshold-based load detection algorithm (classical Markov), the load detection algorithm based on the proposed model (Kth-Order combined Markov model, $K = 1, 2, 3$). For K-order mixed Markov models above 3 orders, due to the sharp rise in its state space (for this experiment, its state space reaches 10,000 and the transition probability matrix is 10,000 steps), the increase in accuracy is not enough to compensate for the complexity. Therefore, this experiment did not test it.

Virtual Machine Migration Quantity Analysis In the simulation experiment, a threshold-based load detection algorithm (classical Markov), a local regression robust (LRR) algorithm, and a K-sequence based hybrid Marker are first tested. The total number of virtual migrations at different times for the Kth-order combined Markov model (Kth-order combined Markov model) is shown in Fig. 2.

From Fig. 2, the effectiveness of the algorithm based on the K-order mixed Markov model (1st-order, 2nd-order, and 3rd-order combined Markov model) is much better than other algorithms, because this algorithm improves the accuracy of the prediction. The rate reduces the probability of allocating virtual machines on hosts that are about to enter high loads, thereby effectively reducing the number of virtual machine migrations. In this experiment, only 1 day of CPU data was used. Since the data space is small, the state space of the K-order mixed Markov model is often relatively large. The probability of occurrence of each state is extremely low, and the overall state coverage of the model is low, leading to 3rd-order combined Markov model and 2nd-order combined Markov model compared to 1st-order combined Markov model The Model) algorithm does not necessarily reduce the number of virtual machine migrations. However, from the results of the various time periods and the final results, the model presented in this paper has effectively reduced the number of virtual machine migrations in the cloud data center.

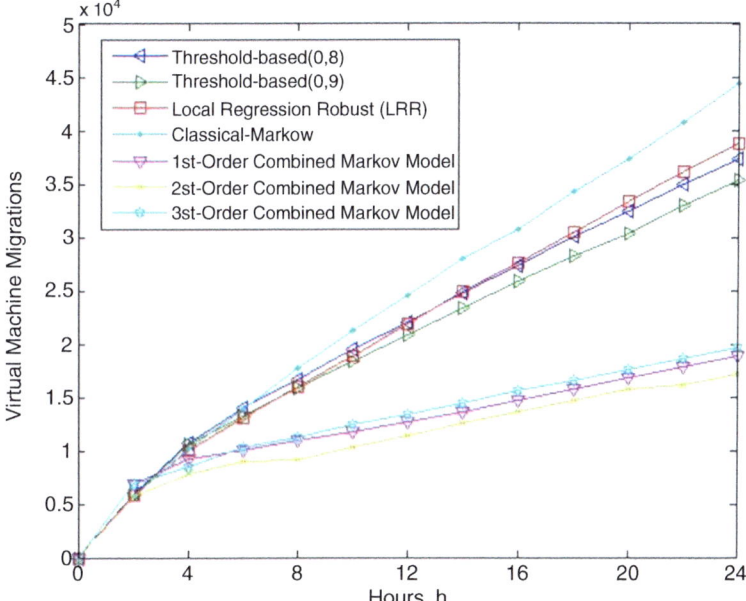

Fig. 2 Comparison of virtual machine migration times

Energy Analysis In the simulation experiment, we continue to test the comparison of each algorithm's energy consumption in the cloud data center. Previously it has been described that the host's consumption of electrical energy is linearly related to the CPU utilization. Therefore, the CPU utilization of each host in the cloud data center can be used to estimate the power of the host and further estimate the energy consumption of the cloud data center. The SPEC (Standard Performance Evaluation Corporation) project lists the relationship between the power consumption of multiple models and configured hosts and the CPU usage, and the same as the query of the project's documentation, you can find the power consumption changes of various host configurations.

The overall energy consumption test results of various algorithms at different times are shown in Fig. 3.

From Fig. 3, we can see that the algorithm proposed in this paper can effectively reduce data center energy consumption. If you frequently allocate virtual machines on hosts with high loads in a short period of time, it is bound to cause frequent migration of virtual machines and increase energy consumption. However, the number of virtual machine migrations is not absolutely related to the energy consumption in the cloud data center because the number of virtual machine migrations is increased when the virtual machine is migrated away from a host with low load, and the low-load host is switched to a low-power state. This will reduce

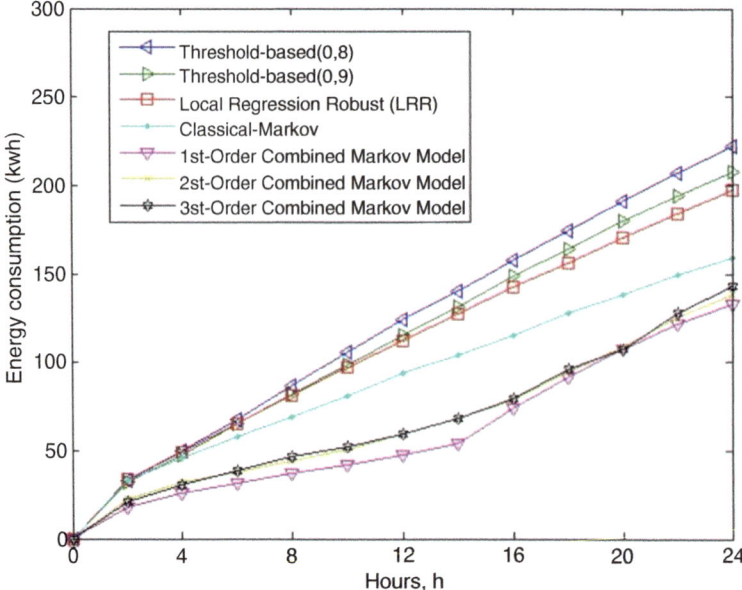

Fig. 3 Energy consumption comparison chart

the power consumption of cloud data centers. The model proposed in this paper can predict the probability that the CPU will enter a high load at a certain moment, and it can avoid the possibility of assigning virtual machines on hosts that are about to enter a high load and avoid migrating virtual machines to hosts that are about to enter a high load. Reduce the overall energy consumption of the data center.

SLA and SLAV Analysis For cloud computing platforms, meeting the requirements of quality of service (QoS) is extremely important. If the algorithm is only blindly pursued to reduce the number of virtual machine migrations and reduce power consumption without considering QoS, the overall performance of the data center may be greatly affected. QoS is usually expressed in the form of SLAs (service level agreement), which is a value that can be measured by the minimum throughput or maximum response time of the system. In our experiments, SLAs can be defined in this way: When the resources requested by an application on a virtual machine can be obtained at 100% at any time, and only limited by the parameter settings when the virtual machine was created, it can be considered that the SLA requirement is satisfied. SLAV (service level agreement violation) refers to the violation of the SLA. Each time a virtual machine is requested and allocated, if the virtual machine's requested computing resources cannot be fully satisfied, that is, the SLA is violated, the measurement can be performed using Eq. (5.1):

$$\text{Overall\_SLAV} = \frac{\sum_{i=1}^{n} (r_i \times t_i) - \sum_{i=1}^{n} (a_i \times t_i)}{\sum_{i=1}^{n} (r_i \times t_i)},$$

$$\text{Average\_SLAV} = \frac{1}{m} \sum_{i=1}^{n} \frac{r_i - a_i}{r_i}$$

(5.1)

Among them, Ovl_SLAV refers to the overall SLA violation, r_i is the resource requested by virtual machine i, a_i is the resource allocated by the virtual machine, Δt_i refers to the time interval between the allocation of virtual machine i and the allocation of the last virtual machine, Average_SLAV is the average SLA in violation, r_i, a_i has the same meaning as Average_SLAV, and m is the number of virtual machines that violated the SLA when requesting and distributing. This experiment considers ample resources such as memory bandwidth and uses CPU resources instead. n is the number of virtual machines.

The comparison of the seven model algorithms is shown in Fig. 4.

The threshold-based load detection algorithms (threshold-based (0.8) and threshold-based (0.9)) and the LRR algorithm have very low overall-SLAV. For these two algorithms, each time the host is determined to enter a high-load state virtual machine migration is performed so that the entire data center hardly has

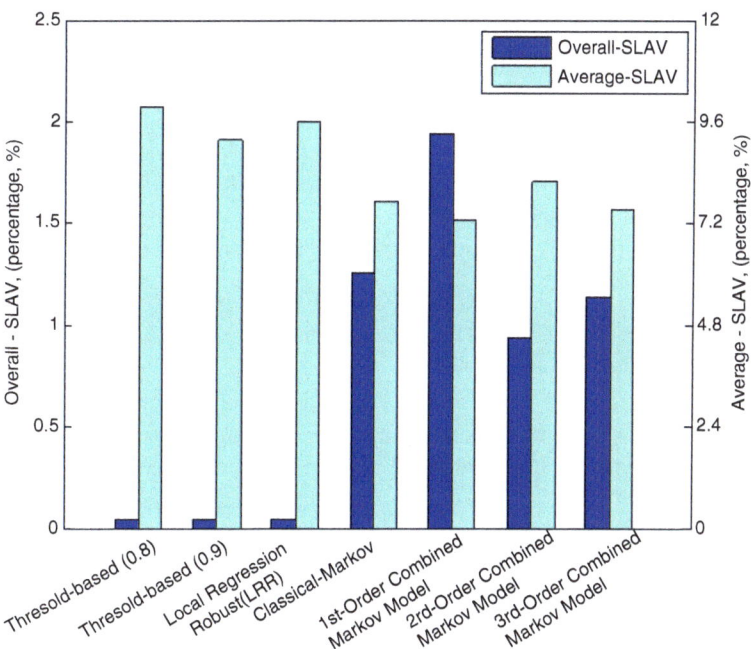

Fig. 4 SLA violates comparison chart

hosts that remain in a high-load state. This virtual machine migration triggering strategy has great benefits for complying with SLAs, but may cause unnecessary virtual machine migration hosting. It only takes a very short period of time to enter a high load, and then it immediately exits from a high-load state, without the need to migrate virtual machines and increase energy consumption. However, the prediction model algorithm proposed in this paper does not guarantee that the prediction result is accurate. Therefore, when the host enters a high-load state and is not predicted, if a virtual machine is requested to be allocated on it, it is bound to violate the SLA. The SLA violation (Overall$_{SLAV}$) is higher than the ordinary non-prediction model algorithm. For the average SLA violation (Average$_{SLAV}$), which measures the average SLA violation, for the current research, when the virtual machine applies for resources, the data center host cannot guarantee that it will be 100% satisfied every time. It does not enter a high-load state, but its remaining resources cannot meet the demand for the required resources when the virtual machine is created. In general, the SLA violation rate of the algorithm based on the K-order mixed Markov model proposed by this paper is higher than other algorithms, but it is in an acceptable range.

6 Conclusion

This paper studies the time-series prediction model and comprehensively considers the predicting effect of multi-order Markov model and the influence of CPU state weights at different moments. A Kth-sequence Markov model is proposed, which also tests the prediction results to reduce the probability of false predictions as much as possible. The experimental comparison shows that the trigger strategy based on the K-order hybrid Markov forecast model proposed in this paper has greater advantages than the traditional model trigger strategy. It can effectively reduce the number of virtual machine migrations and the energy consumption of cloud data centers, and only the SLA violation of the cloud data center system will be lower.

References

1. Zhao, H., & Zhao, J. (2014). Application and analysis of cloud computing technology in digital library. *Library and Information Guide, 24*(7), 33–34.
2. Barroso, L. A. & Hlzle, U. (2007). The case for energy-proportional computing. *Computer, 40*(12), 33–37.
3. Koomey, J. (2011). *Growth in data center electricity use 2005 to 2010* (pp. 41–50). Berkeley: Analytics Press.
4. Lu, H., Li, Y., Mu, S., Wang, D., Kim, H., & Serikawa, S. (2017). Motor anomaly detection for unmanned aerial vehicles using reinforcement learning. *IEEE Internet of Things Journal*(99), 1–1.

5. Lu, H., Li, Y., Chen, M., Kim, H., & Serikawa, H. (2018). Brain intelligence: Go beyond artificial intelligence. *Mobile Networks and Applications, 23*(2), 368–375.
6. Wang, Q., Xiong, W., Zhang, Y., Pan, N., Yu, Z., Song, E., et al. (2018). Remote analysis of myocardial fiber information in vivo assisted by cloud computing. *Future Generation Computer Systems, 85,* 146–159.
7. Zhang, Y., Gravina, R., Lu, H., Villari, M., & Fortino, G. (2018) PEA: Parallel electrocardiogram-based authentication for smart healthcare systems. *Journal of Network and Computer Applications, 117,* 10–16.
8. Xiao, S., Yu, H., Wu, Y., Peng, Z., & Zhang, Y. (2017). Self-evolving trading strategy integrating internet of things and big data. *IEEE Internet of Things Journal, 5*(4), 2518–2525. http://dx.doi.org/10.1109/JIOT.2017.2764957.
9. Zhang, Y., Yang, F., Wang, Q., He, Q., Li, J., & Yang, Y. (2017). An anti-collision algorithm for RFID-based robots based on dynamic grouping binary trees. *Computers & Electrical Engineering, 63,* 91–98. http://www.sciencedirect.com/science/article/pii/S0045790617305098, http://dx.doi.org/https://doi.org/10.1016/j.compeleceng.2017.03.003.
10. Serikawa, S., & Lu, H. (2014). *Underwater image dehazing using joint trilateral filter.* Oxford, Pergamon Press, Inc.
11. Lu, H., Li, Y., Uemura, T., Kim, H., & Serikawa, S. (2018). Low illumination underwater light field images reconstruction using deep convolutional neural networks. *Future Generation Computer Systems, 82,* 142–148.
12. Lu, H., Li, B., Zhu, J., Li, Y., Li, Y., Xu, X., et al. (2017). Wound intensity correction and segmentation with convolutional neural networks. *Concurrency and Computation Practice and Experience, 29*(6), e3927.
13. Xu, X., He, L., Lu, H., Gao, L., & Ji, Y. (2018). Deep adversarial metric learning for cross-modal retrieval. *World Wide Web-internet & Web Information Systems,* 1–16.
14. Calheiros, R. N., Ranjan, R., Beloglazov, A., Rose, C. A. F. D., & Buyya, R. (2010). CloudSim: a toolkit for modeling and simulation of cloud computing environments and evaluation of resource provisioning algorithms, software: Practice and experience. *Software Practice and Experience, 41*(1), 23–50.
15. Beloglazov, A., & Buyya, R. (2012). Optimal online deterministic algorithms and adaptive heuristics for energy and performance efficient dynamic consolidation of virtual machines in cloud data centers. *Concurrency and Computation Practice and Experience, 24*(13), 1397–1420.
16. Li, M. F., Bi, J. P., & Li, Z. C. (2014). Resource scheduling waits for cost-aware virtual machine integration. *Journal of Software, 21*(7), 1388–1402.
17. Hermenier, F., Lorca, X., Menaud, J. M., Muller, G., & Lawall, J. (2009). Entropy: a consolidation manager for clusters. In *ACM SIGPLAN/SIGOPS International Conference on Virtual Execution Environments* (pp. 41–50). Washington, ACM.
18. Verma, A., Ahuja, P., & Neogi, A. (2008). *pMapper: power and migration cost aware application placement in virtualized systems.* Berlin, Springer.
19. Nathuji, R., & Schwan, K. (2007). VirtualPower: coordinated power management in virtualized enterprise systems. *ACM SIGOPS Operating Systems Review, 41*(6), 265–278.
20. Beloglazov, A., & Buyya, R. (2013). Managing overloaded hosts for dynamic consolidation of virtual machines in cloud data centers under quality of service constraints. *IEEE Transactions on Parallel and Distributed Systems, 24*(7), 1366–1379.
21. Wood, T., Shenoy, P., Venkataramani, A., & Yousif, M. (2009). Black-box and gray-box strategies for virtual machine migration. In *Proceedings of the 4th USENIX Conference on Networked Systems Design and Implementation* (pp. 17–17). Berkeley, CA: USENIX Association.
22. Zhu, X., Young, D., Watson, B.J., Wang, Z., Rolia, J., Singhal, S., et al. (2008). 1000 islands: Integrated capacity and workload management for the next generation data center. In *International conference on autonomic computing* (pp. 172–181). Piscataway: IEEE.
23. Gmach, D., Rolia, J., Cherkasova, L., Belrose, G., Turicchi, T., & Kemper, A. (2009). An integrated approach to resource pool management: Policies, efficiency and quality metrics. In

IEEE International Conference on Dependable Systems and Networks with FTCS and DCC (pp. 326–335). Piscataway: IEEE.

24. Gmach, D., Rolia, J., Cherkasova, L., & Kemper, A. (2009). Resource pool management: Reactive versus proactive or let's be friends. *Computer Networks, 53*(17), 2905–2922.

25. Verma, A., Dasgupta, G., Nayak, T. K., De, P., & Kothari, R. (2009). Server workload analysis for power minimization using consolidation. In *Conference on USENIX Technical Conference* (pp. 28–28). Berkeley, CA: USENIX Association.

26. Weng, C., Li, M., Wang, Z., & Lu, X. (2009). Automatic performance tuning for the virtualized cluster system. In *IEEE International Conference on Distributed Computing Systems* (pp. 183–190). Piscataway: IEEE.

27. Bobroff, N., Kochut, A., & Beaty, K. (2007). Dynamic placement of virtual machines for managing SLA violations. In *IFIP/IEEE International Symposium on Integrated Network Management* (pp. 119–128). Piscataway: IEEE.

28. Huang, Q., Shuang, K., Xu, P., Li, J., Liu, X., & Su, S. (2014). Prediction-based dynamic resource scheduling for virtualized cloud systems. *Journal of Networks, 9*(2), 375–383.

29. Beloglazov, A. (2013). Energy-efficient management of virtual machines in data centers for cloud computing. Department of Computing & Information Systems. The University of Melbourne.

30. Khalil, F., Li, J., & Wang, H. (2006). A framework of combining Markov model with association rules for predicting web page accesses. In *Australasian Conference on Data Mining and Analytics* (pp. 177–184). Darlinghurst: Australian Computer Society, Inc.

31. Deshpande, M., & Karypis, G. (2001). Selective Markov models for predicting web page accesses. *ACM Transactions on Internet Technology, 4*(2), 163–184.

32. Xia, L. T. (2005). Prediction of plum rain intensity based on index weighted Markov chain. *Journal of Hydraulic Engineering, 36*(8), 988–993.

33. Peng, Z. (2010). Weighted Markov chains for forecasting and analysis in incidence of infectious diseases in Jiangsu province, China. *The Journal of Biomedical Research, 24*(3), 207–214.

34. Park, K. S., & Pai, V. S. (2006). CoMon: a mostly-scalable monitoring system for PlanetLab. *ACM SIGOPS Operating Systems Review, 40*(1), 65–74.

Multi-Level Chaotic Maps for 3D Textured Model Encryption

Xin Jin, Shuyun Zhu, Le Wu, Geng Zhao, Xiaodong Li, Quan Zhou, and Huimin Lu

1 Introduction

Nowadays, more and more images and videos are flooded in our daily lives. In addition to images and videos, 3D models are beginning to use 3D modeling and 3D printing. Certain apps on smartphones, such as Autodesk 123D Catch, let users to take a themed photo from various views and upload all the photos to the Autodesk cloud server. The 123D service on the cloud server will then return the 3D model of the theme to the user. Desktop software like Google SketchUp can also easily edit 3D models. The 3D models gradually enter our daily life.

Virtual reality technology has become an increasingly popular topic in the industry, and a lot of 3D models are needed to build virtual worlds. The virtual

X. Jin
Department of Cyber Security, Beijing Electronic Science and Technology Institute, Beijing, China

CETC Big Data Research Institute Co., Ltd., Guiyang, Guizhou, China

S. Zhu · L. Wu · G. Zhao · X. Li (✉)
Department of Cyber Security, Beijing Electronic Science and Technology Institute, Beijing, China
e-mail: lxd@besti.edu.cn

Q. Zhou (✉)
National Engineering Research Center of Communications and Networking, Nanjing University of Posts and Telecommunications, Nanjing, P. R. China

State Key Laboratory for Novel Software Technology, Nanjing University, Nanjing, P. R. China
e-mail: quan.zhou@njupt.edu.cn

H. Lu
Department of Mechanical and Control Engineering, Kyushu Institute of Technology, Kitakyushu, Japan

© Springer Nature Switzerland AG 2020
H. Lu, L. Yujie (eds.), *2nd EAI International Conference on Robotic Sensor Networks*, EAI/Springer Innovations in Communication and Computing, https://doi.org/10.1007/978-3-030-17763-8_10

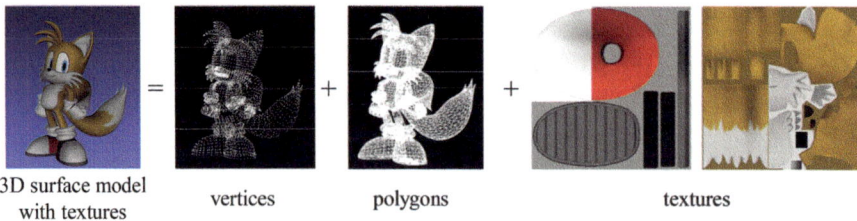

3D surface model
with textures   vertices   polygons   textures

Fig. 1 The basic composition of a 3D surface model consists of vertices, polygons, and textures

reality and augmented reality market is expected to reach $1.06 billion in 2018, with a compound annual growth rate (CAGR) of 15.18% from 2013 to 2018. The government is scanning the entire city's 3D virtual city model with laser scanners and multi-view cameras. In order to achieve a high level of security, integrity, confidentiality, and protection against unauthorized access to sensitive information, 3D content encryption technology is needed. The 3D content is stored or transmitted through unsafe channels.

The 3D digitized objects are defined by two types of 3D content: 3D solid model and 3D surface (shell/boundary) model. The solid model defines the volume of the physical object represented, while the surface model represents the surface, not the volume. Rey solves the encryption problem of 3D entity model in [11].

This paper focuses on the encryption of three-dimensional surface models with texture. The direction of the current method is usually considered point cloud encryption [5, 9], grid [2] and textures [10]. Complete 3D surface models usually contain vertices, polygons, and textures, as shown in Fig. 1.

We have the observation that the vertices, the polygons, and the textures make different levels of contributions to recognize a 3D surface model.

Therefore, in this work, we propose a 3D texture model encryption method based on multi-level chaotic mapping of vertices (point clouds), polygons, and textures. For vertices mainly used for identification, we use advanced 3D Lu chaotic mapping to encrypt them. For polygons and textures with relatively small contributions to recognition, we used cat map of 2D Arnold and 1D logistic map to encrypt them, respectively. The experimental results show that our method achieves similar performance in the same high-level chaotic mapping of vertices, polygons, and textures with less time consuming. In addition, our method can resist many kinds of attacks, such as brute-force attack, statistical attack, and related attack.

2   Previous Work

The special properties of chaos [13], such as sensitivity, pseudorandomness, and ergodicity to initial conditions and system parameters, make chaos dynamics a promising alternative to traditional encryption algorithms. Intrinsic attributes directly relate them to obfuscated and diffuse cryptographic characteristics, which are referenced in Shannon's work [12].

In this respect, the current approach only considers solid models [11], point cloud models [5, 9], meshes [2], and textures [10]. This paper examines the texture encryption of the most complex 3D surface models.

To the best of our knowledge, the most similar work to ours is Jin et al. [7], in which they also propose a method for the encryption of 3D surface models with textures. However, they use the same high-level chaotic maps for vertices, polygons, and textures. This makes the time consumption of their method is high. Based on our observations in Sect. 1, we choose to use a multi-level (hierarchical) way of using high-level chaotic for vertices, middle one for polygons, and low one for textures.

3 Preliminaries

The three levels of chaotic maps we leveraged in this work are 1D logistic map, 2D Arnold's cat map, and 3D Lu map. Because of its high complexity, high-dimensional chaotic systems are more reliable in designing secure image encryption schemes. The cryptosystem based on low-dimensional chaotic mapping has shortcomings such as short cycle and small key space [3, 4, 6, 8, 14]. However, the low-dimensional chaotic maps require less computational cost than that of the high-dimensional chaotic maps. Thus in this paper, we combine the high security of high-dimensional chaotic maps and the high speed of low-dimensional chaotic maps.

3.1 1D Logistic Map

The basic formula of 1D chaotic encryption is as follows:

$$x_{n+1} = \mu x_n (1 - x_n)$$
$$3.569945672\ldots < \mu \leq 4, 0 \leq x_n \leq 1 \tag{1}$$
$$n = 0, 1, 2, \ldots.$$

When $3.569945672\ldots < \mu \leq 4, 0 \leq x_0 \leq 1$, the system is in chaotic state.

3.2 2D Arnold's Cat Map

Cat mapping is a chaotic system that iterates or evolves the plaintext as the initial value of the chaotic system to achieve the effect of scrambling the plaintext

$$\begin{bmatrix} X' \\ Y' \end{bmatrix} = \begin{bmatrix} 1 & p \\ q & p*q+1 \end{bmatrix} * \begin{bmatrix} X \\ Y \end{bmatrix} \bmod U \tag{2}$$

$$\begin{bmatrix} X \\ Y \end{bmatrix} = \begin{bmatrix} 1 & p \\ q & p*q+1 \end{bmatrix}^{-1} * \begin{bmatrix} X' \\ Y' \end{bmatrix} \bmod U, \tag{3}$$

where p and q represent the positive secret keys. (X, Y) is the original 2D variables. (X', Y') is the new values of (X, Y). U is the upper bounds of values of X and Y.

3.3 3D Lu Map

The Lu map is a 3D chaotic map. It is described by Eq. (4)

$$\begin{cases} \dot{x} = a(y - x) \\ \dot{y} = -xz + cy, \\ \dot{z} = xy - bz \end{cases} \tag{4}$$

where (x, y, z) are the system trace. (a, b, c) are the system parameters. When $a = 36$, $b = 3$, $c = 20$, the system contains a strange attractor and is in chaotic state.

4 Multi-Level 3D Model Encryption

In this section, we describe the proposed encryption method for a textured 3D surface model. First, the 3D texture model is decomposed into vertices, polygons, and textures. Then, the three parts are encrypted by using three-dimensional Lu map, two-dimensional Arnold graph, and one-dimensional logistic graph, respectively, as described in Sect. 3. Finally, the encrypted vertices, polygons, and textures are combined into an encrypted 3D texture model (Fig. 2).

4.1 Vertices Encryption

The vertices in the 3D texture model are expressed by the following triple list:

$$V = \{(X_1, Y_1, Z_1), \ldots, (X_N, Y_N, Z_N)\}, \tag{5}$$

where (X_i, Y_i, Z_i) is the 3D coordinate of a vertex. N is the number of the vertices. We use the 3D Lu map defined in Eq. (4) to produce a random vector with dimensions of $3N$:

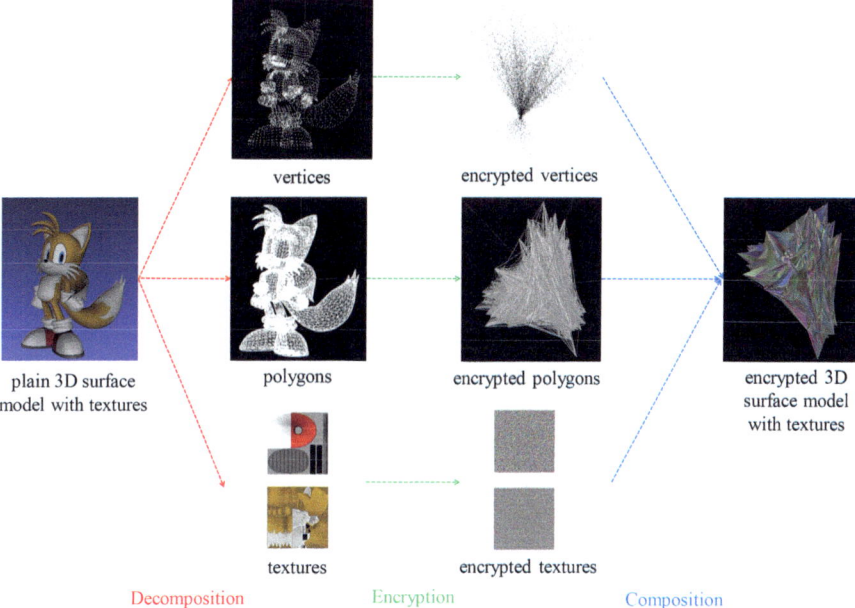

Fig. 2 Our presented 3-dim textured encryption model. The decryption method is the inverse version of the encryption method

$$LV = \{(LV_1, LV_2, LV_3), \ldots, (LV_{3N-2}, LV_{3N-1}, LV_{3N})\}. \tag{6}$$

Then we make element by element product of V and LV:

$$VLV =$$
$$\{(X_1 LV_1, Y_1 LV_2, Z_1 LV_3), \ldots, (X_N LV_{3N-2}, Y_N LV_{3N-1}, Z_N LV_{3N})\}. \tag{7}$$

The new vector VLV contains the new coordinates of the original 3D vertex:

$$(X_i, Y_i, Z_i) \rightarrow$$
$$(X_i LV_{3(i-1)}, Y_i LV_{3(i-1)+1}, Z_i LV_{3(i-1)+2}), 1 \leq i \leq N. \tag{8}$$

4.2 Polygons Encryption

The polygons (taking the triangle as an example) in a 3D textured model are in the form of a list of triplets:

$$P = \{(A_1, B_1, C_1), \ldots, (A_i, B_i, C_i), \ldots, (A_M, B_M, C_M)\}, \tag{9}$$

where (A_i, B_i, C_i) represents the 3 vertices of a triangle in the form of the indices of vertices. $1 \leq i \leq M, 1 \leq A_i, B_i, C_i \leq N$. N is the number of the vertices. We use the 2D Arnold's Cat map defined in Eqs. (2) and (3) to produce a random vector with dimensions of $3M$:

$$LP = \{(LP_1, LP_2, LP_3), \dots, (LP_{3M-2}, LP_{3M-1}, LP_{3M})\}. \tag{10}$$

We make the element-to-element correspondences between P and LP:

$$\begin{cases} A_i \longleftrightarrow LP_{3(i-1)+1} \\ B_i \longleftrightarrow LP_{3(i-1)+2} \\ C_i \longleftrightarrow LP_{3(i-1)+3}. \end{cases} \tag{11}$$

Then we make ascending sort of LP. The sorted LP is denoted as LP^{sort}. According to the new order in LP^{sort}, we reorder the element in P using the correspondences described in Eq. (11). The vector with new order of P is denoted as P':

$$P' = \{(A_1', B_1', C_1'), \dots, (A_i', B_i', C_i'), \dots, (A_M', B_M', C_M')\}, \tag{12}$$

where (A_i', B_i', C_i') is the new triangle of the encrypted 3D model.

4.3 Textures Encryption

In texture 3D model, texture is represented as 2D image with corresponding texture coordinates. We use 1D logic mapping based image encryption method and DNA coding [4] to encrypt the texture image. We first divide the texture image into RGB channels. Each channel of the texture image is then encoded by DNA coding. We then used the 1D logical mapping to generate a random matrix of texture images of the same size, and added it to the coding results using DNA additions. After that, another random matrix with the same size texture image is generated by 1D logical mapping and converted into a binary matrix with a threshold of 0.5. Then, when the corresponding value in the second random matrix is 1, the DNA addition result is converted to the DNA complement result. The final step is to decode the DNA to get an 8-bit encryption result.

5 Simulation Performance

We use plenty of 3D textured models to test our method, as shown in Fig. 3, with the secret keys shown in Table 1.

| plain clock | cipher clock | plain dog | cipher dog |

Fig. 3 The simulation results. We test our method on 3D models with various contents

Table 1 The secret keys of the 3D Lu maps in Eq. (4)

Encryption phases	Keys
Vertices encryption	$x_0^v = -6.045$, $y_0^v = 2.668$, $z_0^v = 16.363$
Polygons encryption	$p = 1, q = 1$
Texture encryption	$x_0^{t1} = 0.62$, $\mu^{t1} = 3.99$, $x_0^{t2} = 0.26$, $\mu^{t2} = 3.9$

In the vertices encryption phases, $a = 36, b = 3, c = 20$

In our approach, we use 1 Lu map, 1 Arnold's cat map, and 2 logistic maps. Texture encryption contains 2 logistic maps. Three-dimensional texture models with different content were tested. All encryption results can be correctly decrypted to the original pure 3D model with the correct secret key. The results show that the simulation results are satisfactory.

6 Security and Performance Analysis

Well-designed 3D model encryption schemes should be able to resist various attacks, such as violent attacks and statistical attacks. In this section, we will analyze the security of the proposed encryption method.

6.1 Resistance to the Brute-Force Attack

Key Space The key space of the 3D model encryption scheme should be large enough to resist violent attacks; otherwise, it will be broken down by exhaustive search to obtain the secret key in a limited amount of time. In our encryption method, we have the key space of 9 key values shown in Table 2.

The precision of 64-bit double data is 10^{-15}; thus, the key space is about $(10^{15})^9 = 10^{135} \approx 2^{449}$, which is much larger than the max key space (2^{256}) of practical symmetric encryption of the AES [1]. In our practice, we use double data to simulate large integers for p and q in Table 1. Our key space is large enough to resist brute-force attack.

Table 2 The key spaces

Chaotic maps	Key spaces
3D Lu	$-40 < x_0^v < 50, -100 < y_0^v < 80, 0 < z_0^v < 140$
2D Arnold' cat map	p, q are positive integers
1D logistic map	$3.569945672\ldots < \mu^{t1}, \mu^{t2} \leq 4, x_0^{t1}, x_0^{t2} \in [0,1]$

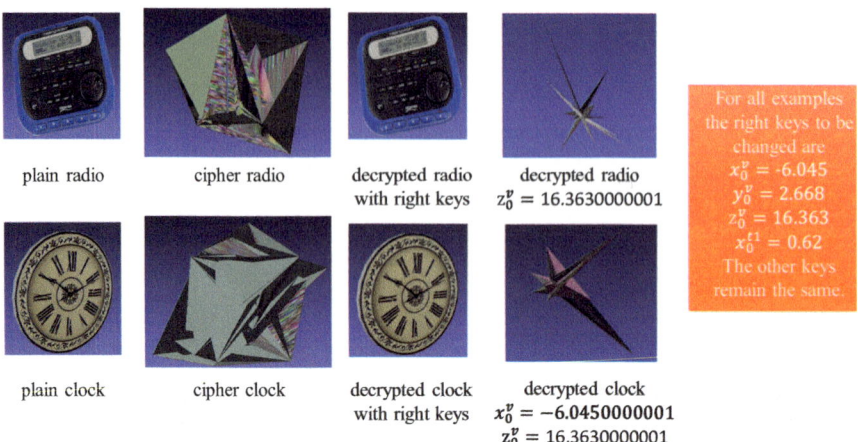

Fig. 4 Decrypted with wrong key. We slightly change the key and get the wrong decrypted result

Sensitivity of Secret Key The chaotic systems are extremely sensitive to the system parameter and initial value. A light difference can lead to the decryption failure. To test the secret key sensitivity of the 3D model encryption scheme.

We use the modified key to decrypt the encrypted 3D surface model, while the other keys remain unchanged. The decryption result is shown in Fig. 4. We can see that the decrypted 3D model is completely different from the original plain 3D model. The test results for the other key are similar. The experiments show that the 3D model encryption scheme is very sensitive to keys and has strong resistance to exhaustive attacks.

6.2 Resistance to the Statistic Attack

The Histogram Analysis For the vertex, the viewpoint feature histogram (VFH) is the representation of point clusters for cluster identification and 6 DOF pose estimation problems. We use VFH to evaluate our 3D vertex encryption. As shown in Fig. 5, the VFH of our method's encryption results is completely different from the VFH of the original 3D model, which makes statistical attacks impossible.

Distribution of Occupied Positions We further analyze the occupied positions of the 3D vertices. As defined in [11], we compute the occupied position per x-column, y-column, and z-column of a 3D lattice $Z = (z_{ijk})$.

The matrix is very different for normal 3D vertices and encrypted 3D vertices. The occupied position of each z-column in the plane 3D vertex and the corresponding encrypted 3D vertex are shown in Fig. 6. The positions of the occupation were distributed widely. In the case of ordinary 3D vertices, some clusters appear, while in the case of encrypted 3D vertices, the distribution appears to be uniform.

6.3   The Speed of the Encryption and Decryption

Our 3D surface model encryption scheme is implemented on personal computers by Matlab with AMD A10 PRO-7800B, 12 computer cores 4C + 8G 3.4GHz, and 4.00g RAM. Time cost encrypting and decrypting 3D models with different number of vertices. The larger the size of the 3D model, the more the time it takes to encrypt and decrypt it. When we implemented the migration to other tool environments (such as C/C++) in Matlab 2015a, the speed was faster and could meet the actual needs.

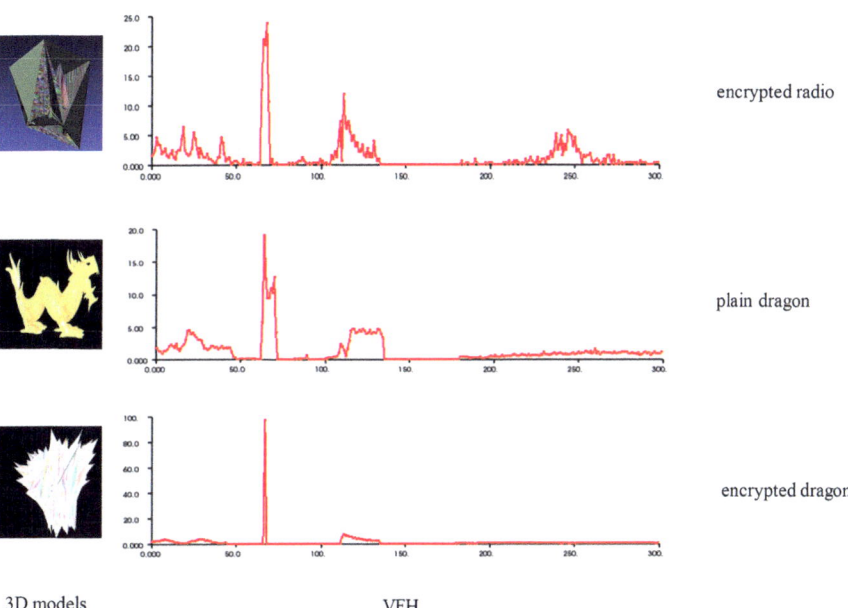

3D models   VFH

Fig. 5 The viewpoint feature histogram (VFH) of 3D textured models before and after encryption

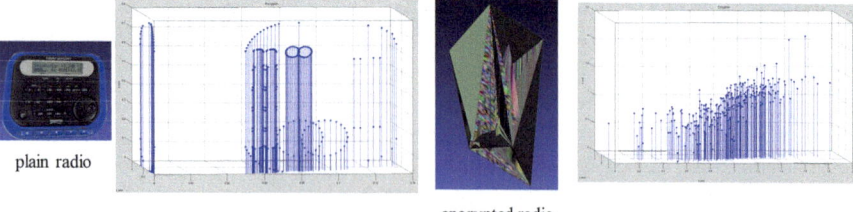

plain radio

encrypted radio

Fig. 6 The distribution of occupied positions per z-column of the 3D textured models before and after encryption

7 Conclusions

In this paper, we propose a hierarchical multi-level encryption scheme based on multiple dimensional chaotic maps for 3D space models with textures. Our method is based on the observations that the vertices make more contribution to recognizing a 3D model than polygons and textures. The hierarchical scheme reduces the time consumption of encryption and decryption compared with the methods of Jin et al. [7], who made equal level encryption on vertices, polygons, and textures, while our encryption results and the key sensitivity are nearly the same to those of [7]. In the future work, we will try to find more hierarchical encryption schemes for 3D models and rely on more sophisticated psychological experiments on 3D model recognition.

Acknowledgements This work is partially supported by the National Natural Science Foundation of China (Grant Nos. 61772047, 61772513), the Science and Technology Project of the State Archives Administrator (Grant No. 2015-B-10), and the Fundamental Research Funds for the Central Universities (Grant No. 328201803, 328201801).

References

1. Bogdanov, A., Khovratovich, D., & Rechberger, C. (2011). Biclique cryptanalysis of the full AES. In *Proceedings of the 17th International Conference on the Theory and Application of Cryptology and Information Security* (pp. 344–371). Berlin: Springer.
2. Éluard, M., Maetz, Y., & Doërr, G. (2013). Geometry-preserving encryption for 3D meshes. In *Actes de Compression et REprsentation des Signaux Audiovisuels* (pp. 7–12).
3. Jin, X., Chen, Y., Ge, S., Zhang, K., Li, X., Li, Y., et al. (2015). Color image encryption in CIE L*a*b* space. In *International Conference on Applications and Techniques in Information Security* (pp. 74–85). Berlin: Springer.
4. Jin, X., Tian, Y., Song, C., Wei, G., Li, X., Zhao, G., et al. (2015). An invertible and anti-chosen plaintext attack image encryption method based on DNA encoding and chaotic mapping. In *2015 Chinese Automation Congress (CAC)* (pp. 1159–1164). Piscataway: IEEE.
5. Jin, X., Wu, Z., Song, C., Zhang, C., & Li, X. (2016). 3D point cloud encryption through chaotic mapping. In *Advances in Multimedia Information Processing (PCM) 2016. Proceedings of the 17th Pacific-rim Conference on Multimedia* (pp. 119–129). Cham: Springer.

6. Jin, X., Yin, S., Li, X., Zhao, G., Tian, Z., Sun, N., et al. (2016). Color image encryption in YCbCr space. In *8th International Conference on Wireless Communications & Signal Processing, WCSP* (pp. 1–5). Piscataway: IEEE.
7. Jin, X., Zhu, S., Xiao, C., Sun, H., Li, X., Zhao, G., et al. (2017). 3D textured model encryption via 3D Lu chaotic mapping. *Science China Information Sciences, 60,* 122107.
8. Jin, X., Yin, S., Liu, N., Li, X., Zhao, G., & Ge, S. (2018). Color image encryption in non-RGB color spaces. *Multimedia Tools and Applications, 77,* 15851–15873.
9. Jolfaei, A., Wu, X., & Muthukkumarasamy, V. (2015). A 3D object encryption scheme which maintains dimensional and spatial stability. *IEEE Transactions on Information Forensics and Security, 10*(2), 409–422.
10. Jolfaei, A., Wu, X., & Muthukkumarasamy, V. (2016). A secure lightweight texture encryption scheme. In *Image and Video Technology – PSIVT 2015 Workshops. PSIVT 2015* (pp. 344–356). Cham: Springer
11. del Rey, Á. M. (2015). A method to encrypt 3D solid objects based on three-dimensional cellular automata. In *Hybrid Artificial Intelligent Systems. 10th International Conference on Hybrid Artificial Intelligence Systems* (pp. 427–438). Cham: Springer
12. Shannon, C. (1949). Communication theory of secrecy systems. *Bell System Technical Journal, 28,* 656–715.
13. Verma, O. P., Nizam, M., & Ahmad, M. (2013). Modified multi-chaotic systems that are based on pixel shuffle for image encryption. *Journal of Information Processing Systems, 9*(2), 271–286.
14. Zhen, P., Zhao, G., Min, L., & Jin, X. (2016). Chaos-based image encryption scheme combining DNA coding and entropy. *Multimedia Tools and Applications, 75*(11), 6303–6319.

Blind Face Retrieval for Mobile Users

Xin Jin, Shiming Ge, Chenggen Song, Le Wu, and Hongbo Sun

1 Introduction

In today's mobile Internet era, increasing mobile users backup their photos to the cloud storage servers. Some cloud servers provide face retrieval service, which allows one to retrieve photos that contain a specific person or a group of persons from all his/her photos in the own storage space of the cloud server.

Furthermore, as shown in Fig. 1, people in one team or one family may share the same cloud storage space and upload their photos together. For example, people in one family take photos of each other using their mobile phones for a long time. In the traditional way, they may copy photos in each mobile phone using a cable and manage their photos manually. Nowadays, one can create a cloud storage space and share it to all the family members. All the photos shot by the family members can be stored in the shared cloud storage space automatically by the cloud Apps in their mobile phones. After that, each family member can browse photos using the cloud Apps in a friendly way. Besides, one can retrieve the photos that contain one or multiple specific family members using the photo management module of the cloud Apps, as shown in Fig. 1.

X. Jin
Department of Cyber Security, Beijing Electronic Science and Technology Institute, Beijing, China

CETC Big Data Research Institute Co., Ltd., Guiyang, Guizhou, China

S. Ge (✉)
Institute of Information Engineering, Chinese Academy of Sciences, Beijing, China
e-mail: geshiming@iie.ac.cn

C. Song · L. Wu · H. Sun
Beijing Electronic Science and Technology Institute, Beijing, China

© Springer Nature Switzerland AG 2020
H. Lu, L. Yujie (eds.), *2nd EAI International Conference on Robotic Sensor Networks*, EAI/Springer Innovations in Communication and Computing, https://doi.org/10.1007/978-3-030-17763-8_11

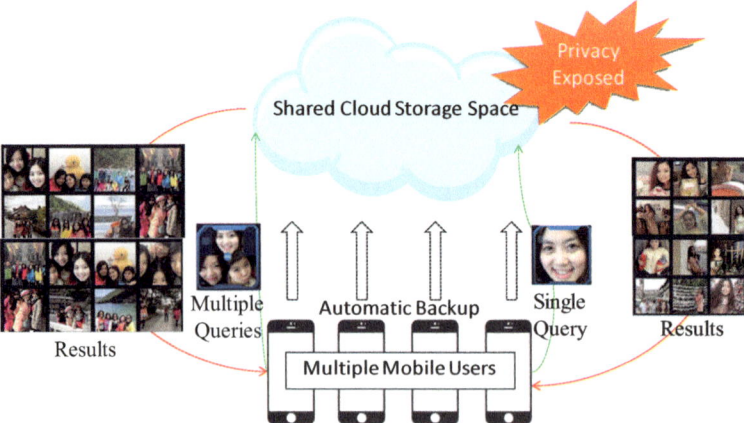

Fig. 1 The typical scenario. Multiple users backup and share their photos in their mobile phones to a cloud storage. However, the privacy of photos is completely exposed to the cloud

The above application models are typical scenarios in today's mobile Internet era. However, although the management of group photos is much more convenient, the privacy of the users' photos is completely exposed to the cloud server. The facial features of each member in one family, the relationship between members, and the school of their children, etc. can be learned from the family photos, which can threaten the personal and property security of the family.

In the meanwhile, the face detector used in the face retrieval task may be trained in a large scale of face images annotated by thousands of people. The copyright of the trained parameters of the face detector should also be preserved from the commercial provider's perspective.

Thus, in this paper, we propose a novel protocol to preserve the privacy of the cloud users' photos and the parameters of the commercial face detector simultaneously in such mobile cloud scenarios. The face retrieval problem can be decomposed into *face detection*, *face recognition*, and *face label matching*. In the face detection stage, face regions are detected in users' photos with rectangles. In the recognition stage, each detected face is marked by a label of a member in a group. Then a label vector is generated according to the face recognition result for each photo so as to mark who is/are in each photo. The above is the off-line phase. In the on-line phase, a user queries a specific face of one person or faces of a group of person. Then, a label vector is generated for this query and compared to each label vector corresponding to each photo. Photos with the most similar label vector to the query label vector are selected as the retrieval result.

Related Work The secure face detection method is proposed as Blind Vision [1] for securely evaluating a Viola–Jones type face detector. After that Jin et al. accelerate secure face detector by introducing a random base image representation [9]. A system called secure computation of face identification (SCiFI) [10] is developed for secure face recognition. This system uses two cryptographic tools (homomorphic encryption and oblivious transfer) to implement a privacy preserving computation of the Hamming distance between two binary vectors. Recently, a lot of researchers have addressed the privacy preserving computer vision problems [2, 3, 5, 7, 8, 10–13]. Most of them leverage the cryptography tools which are not efficient. The main mechanism in our protocols is to security compute the inner product. In 2009, Wong et al. [16] proposed a secure kNN (k-nearest neighbor) scheme on encrypted database, which developed a new asymmetric encryption that preserves inner product. We tailor the encryption scheme to meet our scenario, and construct our privacy preserving face retrieval application.

Our Approach In this paper, we leverage a simple but efficient secure inner production protocol to protect the contents of user photos and the parameters of the face detector. The cloud server only provides the resources of storage and computing and cannot learn anything from the user photos and the face detector. The face detection stage is protected by the secure face detection protocol using our secure inner production protocol. The face recognition stage is running locally in the users' mobile phone. The photos are encrypted and uploaded to the shared cloud storage space together with the corresponding label vector. The face label matching stage is running in the on-line face retrieval phase. The query label vector and the label vector in the cloud are compared using our secure inner production protocol.

2 Problem Formulation

2.1 Overview

Our proposed methods are shown in Fig. 2. In the face detection stage, face regions are detected in users' photos with a rectangle. In the recognition stage, each detected face is marked by a label of a member in a group. Then a label vector is generated according to the face recognition result for each photo so as to mark who is/are in each photo. The above is the off-line phase. In the on-line phase, a user queries a specific face of one person or faces of a group of person. Then, a label vector is generated for this query and compared to each label vector corresponding to each photo. Photos with the most similar label vector to the query label vector are selected as the retrieval result.

2.2 Security Model

We adopt the "honest-but-curious" model for the cloud server. It assumes that the cloud server would honestly follow the designated protocols and procedures to fulfill its service provider's role, while it may analyze the information stored and processed on the server in order to learn additional information about its customers.

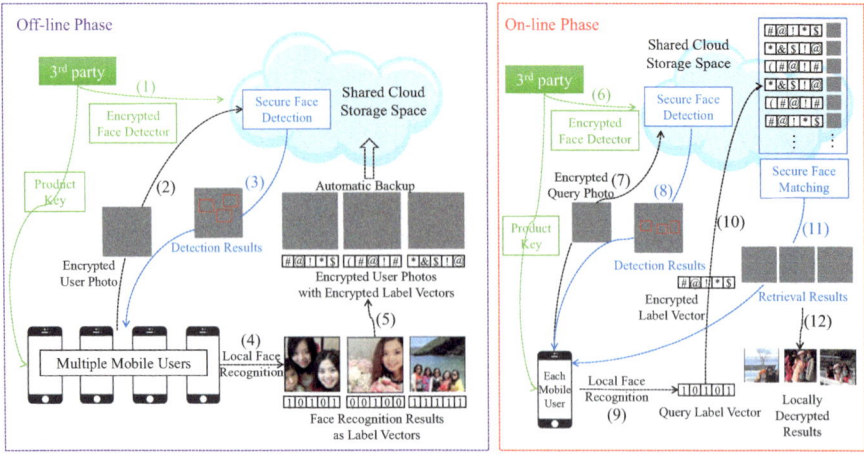

Fig. 2 The overall system architecture. (1) The face detector is provided by a 3rd party with a product key. (2) A user encrypts photos with the same key as used in (1), before sending them to the cloud. (3) Our secure face detector protocol is running in the cloud with the encrypted face detector and photos. The detected face windows are sent back to the user. (4) The local face recognition algorithm is called to mark each photo with a label vector, which reveals who is/are in this photo using 1 for exist. (5) The photos in all the shared users are encrypted and uploaded to the cloud storage together with the encrypted label vectors. The off-line phase is end. In the on-line phase, a user wants to query photos from the cloud storage with all the faces in the query photo. (6) (7) (8) (9) is the same as (1) (2) (3) (4). Then, the label vector of the query photo is computed. (10) The query label vector is encrypted and uploaded to the cloud and compared to all the label vectors in the cloud using our secure face matching protocol. (11) The corresponding encrypted corresponding photos with the top N matching label vectors are sent back to the user. (12) The user decrypts the matching photos and obtains the final retrieval results

The objective of our scheme is to preserve the 3rd party and users' data privacy, which includes: (1) face detector privacy; (2) detected windows privacy; (3) photos content privacy; (4) label vectors privacy; (5) query privacy. While photos content privacy can be achieved by encryption-before-outsourcing schemes, this paper focuses on preserving the data privacy due to the face detection and matching, as follows.

Detection Privacy Besides the detection result, the cloud server should not deduce any face classifier information from the secure face detector, and face information from the secure detected windows.

Matching Privacy Besides the matching result, the cloud server should not deduce any face information from the secure label vectors and secure query.

3   Secure Face Retrieval

3.1   Secure Face Detection

Denote some finite field F that is large enough to represent all the intermediate results. Denote by X the image that Alice owns. A particular detection window within the image X will be denoted by $x \in F^L$ and x will be treated in vector form. Bob owns a strong classifier of the form

$$H(x) = \text{sign}\left(\sum_{n=0}^{N-1} h_n(x)\right), \tag{1}$$

where $h_n(x)$ is a threshold function of the form

$$h_n(x) = \begin{cases} \alpha_n & x^T y_n > \theta_n \\ \beta_n & \text{otherwise,} \end{cases} \tag{2}$$

and $y_n \in F^L$ is the hyperplane of the threshold function $h_n(x)$. The parameters $\alpha_n \in F$, $\beta_n \in F$ and $\theta_n \in F$ of $h_n(x)$ are determined during training; N is the number of weak classifiers used.

As in Fig. 2, step (1), the 3rd party $3P$ first generates the product key according to the users US's purchase as:

D-KeyGen (m)   Given a security parameter m as the most length of the classifiers in the face detector, output the product key $SK(M_1, M_2, S)$, where $M_1, M_2 \in \mathcal{R}^{m \times m}$ are randomly invertible matrices and $S \in \{0, 1\}^m$ is a randomly vector.

the next, $3P$ send this product key to US via secure channel.

The second $3P$ encrypt is classifiers in the face detector and upload to the cloud server with the detect parameter $\{\alpha_i, \beta_i, \theta_i\}_{i=1,...,n}$.

E-FD (SK, Y)   To encrypt the classifiers $Y = y_1, \ldots, y_n$ in the face detector, $3P$ split each vector y_i into two vectors $\{y_i', y_i''\}$ following the rule: for each $y_{i,j} \in y_i$, set $y_{i,j}' = y_{i,j}'' = y_{i,j}$ if $s_j \in S$ is 1; otherwise $y_{i,j}' = \frac{1}{2}y_{i,j} - r$ and $y_{i,j}'' = \frac{1}{2}y_{i,j} + r$ where $r \in \mathcal{R}$ is a random number. Then encrypt $\{y_i', y_i''\}$ with (M_1, M_2) into $\{M_1^T y_i', M_2^T y_i''\}$. Output $EY = \{M_1^T y_i', M_2^T y_i''\}_{i=1,...,n}$

As in Fig. 2, step (2), to detect whether a detection window is a face, US encrypt the window and upload to the cloud server.

E-DW (SK, w)   To encrypt the window w, US split vector w into two vectors $\{w', w''\}$ following the rule: for each $w_j \in w$, set $w_j' = w_j' = w_j$ if $s_j \in S$ is 0; otherwise $w_j' = \frac{1}{2}w_j - r'$ and $w_j'' = \frac{1}{2}w_j + r'$ where $r' \in \mathcal{R}$ is another random number. Then encrypt $\{w', w''\}$ with (M_1, M_2) into $\{M_1^{-1}w', M_2^{-1}w''\}$. Output $EW = \{M_1^{-1}w', M_2^{-1}w''\}$

After receiving the secure classifiers and secure detected window, the cloud server output the detection results as in Fig. 2, step (3).

DC (EY, EW) For each secure classifier $\{M_1^T y_i', M_2^T y_i''\}$, the cloud server first computes

$$t_i = (M_1^T y_i')^T \cdot M_1^{-1}w' + (M_2^T y_i'')^T \cdot M_2^{-1}w'' = y_i^T \cdot w$$

and set $h_i = \alpha_i$ if $t_i \geq \theta_i$ or $h_i = \beta_i$ otherwise. At last, the could server output $H = sign(\sum_{i=1}^{n} h_i)$ as the detection result.

3.2 Face Recognition and Label Vector

After detecting all the face in the photos, as in Fig. 2, step (4), the users US run the face recognition algorithm, i.e., SPR [17], and form the face label vector $L_i \in \{0, 1\}^t$ for each photo, which describes who is/are in each photo. US set the label set $\{L_i\}$ as the index of the photo.

3.3 Secure Face Label Matching

In order to build the secure label vectors of the photo set, the users US first generate the private key as follows:

M-KeyGen (t) Given a security parameter t as the totally face number of the photo set, output the private key $PrK(N_1, N_2, T)$, where $S_1, S_2 \in \mathcal{R}^{t \times t}$ are randomly invertible matrices and $T \in \{0, 1\}^t$ is a randomly vector.

As in Fig. 2, step (5), US encrypt is label vectors and upload to the cloud server.

E-LV (PrK, L) To encrypt the label vector L_i in the vector set, US split each vector L_i into two vectors $\{L_i', L_i''\}$ following the rule: for each $L_{i,j} \in L_i$, set $L_{i,j}' = L_{i,j}'' = L_{i,j}$ if $t_j \in T$ is 1; otherwise $L_{i,j}' = \frac{1}{2}L_{i,j} - u$ and $L_{i,j}'' = \frac{1}{2}L_{i,j} + u$ where $u \in \mathcal{R}$ is a random number. Then encrypt $\{L_i', L_i''\}$ with (N_1, N_2) into $\{N_1^T L_i', N_2^T L_i''\}$. Output $EL = \{N_1^T L_i', N_2^T L_i''\}_{i=1,\cdots,t}$

Next, the users US choose the standard encrypted algorithm such as AES [4], or other photo encryption scheme such as [6], with their own secret key, to encryption the photos, and upload to the cloud server with the secure label vectors.

To search the photos with target faces, US first generate the query as $Q = \{0, 1\}^t$, as in Fig. 2, step (9), where $Q_i = 1$ if the i-th face is one of the target faces, then US encrypt the query and upload it to the cloud server with the amount of target faces λ, as in Fig. 2, step (10).

E-Q (PrK, Q) To encrypt the query Q, US split vector Q into two vectors $\{Q', Q''\}$ following the rule: for each $Q_j \in Q$, set $Q'_j = Q'_j = Q_j$ if $t_j \in T$ is 0; otherwise $Q'_j = \frac{1}{2}Q_j - v$ and $Q''_j = \frac{1}{2}w_j + v$ where $v \in \mathcal{R}$ is another random number. Then encrypt $\{Q', Q''\}$ with (N_1, N_2) into $\{N_1^{-1}Q', N_2^{-1}Q''\}$. Output $EQ = \{N_1^{-1}Q', N_2^{-1}Q''\}$

After receiving the secure index and secure query, the cloud server output the matching result.

MAT (EL, EQ) For each secure label $\{N_1^T L'_i, N_2^T L''_i\}$, the cloud server first computes

$$ret_i = (N_1^T L'_i)^T \cdot N_1^{-1}Q' + (N_2^T L''_i)^T \cdot N_2^{-1}Q'' = L_i^T \cdot Q$$

if $ret_i = \lambda$ the cloud server, then set the i-th photo with the label L_i as one of the matching photos.

As in Fig. 2, step (11), the cloud server return the encryption retrieval photos to US, then US decryption them to get the matching photos.

4 Experiments

We convert the Viola–Jones type face detector [14, 15] to our secure face detector, which is implemented by OpenCV 2.4.3.[1] package. The face detector consists of a cascade of 22 rejectors. The first rejector consists of 3 weak classifiers. The most complicated rejector consists of 213 weak classifiers. There is a total of 2135 weak classifiers.

In this section, we show an experiment on photos from an authorized family, which consists of 5 family members with 4 mobile phones. We use 100 photos (20 photos for each member) to build the dictionary. The number of total family photos in the simulated cloud is 1000. The secure face detection results are shown in Fig. 3.

---

[1]http://opencv.org/.

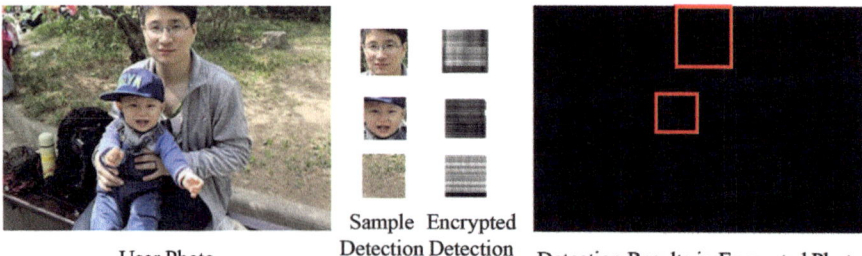

Fig. 3 The secure face detection. The user photo are divided into detection windows, which are sent to the cloud one by one. The detection results are shown in rectangle

5 Conclusion

In this paper, we propose a novel protocol to preserve the privacy of the users' photos and the parameters of the commercial face detector simultaneously in mobile cloud scenarios. The experimental results reveal that our protocol can successfully retrieve the proper photos from the cloud server and protect the user photos and the face detector. One of the cores of the convolutional neural network (CNN) is inner product. Thus, in the future work, we will extend our approach to privacy preserving deep learning framework for face retrieval.

Acknowledgements This work is partially supported by the National Natural Science Foundation of China (Grant Nos. 61772047, 61772513), the Science and Technology Project of the State Archives Administrator (Grant No. 2015-B-10), and the Fundamental Research Funds for the Central Universities (Grant No. 328201803, 328201801).

References

1. Avidan, S., & Butman, M. (2006). Blind vision. In *Proceedings of the 9th European Conference on Computer Vision – Volume Part III* (pp. 1–13). Berlin: Springer.
2. Bost, R., Popa, R. A., Tu, S., & Goldwasser, S. (2015). Machine learning classification over encrypted data. In *22nd Annual Network and Distributed System Security Symposium, NDSS 2015*, San Diego, California, USA, February 8–11, 2014.
3. Chu, C., Jung, J., Liu, Z., & Mahajan, R. (2014). sTRACK: Secure tracking in community surveillance. In *Proceedings of the ACM International Conference on Multimedia, MM'14* (pp. 837–840). New York, NY: ACM.
4. Daemen, J., & Rijmen, V. (2002). *The Design of Rijndael: AES – the Advanced Encryption Standard*. Berlin: Springer.
5. Fanti, G. C., Finiasz, M., & Ramchandran, K. (2013). One-way private media search on public databases: The role of signal processing. *IEEE Signal Processing Magazine, 30*(2), 53–61.
6. Jin, X., Chen, Y., Ge, S., Zhang, K., Li, X., Li, Y., et al. (2015). Applications and techniques in information security. In *Proceedings of the 6th International Conference, ATIS 2015 – Color Image Encryption in CIE L*a*b* Space* (pp. 74–85). Berlin: Springer.

7. Jin, X., Guo, K., Song, C., Li, X., Zhao, G., Luo, J., et al. (2016). Private video foreground extraction through chaotic mapping based encryption in the cloud. In *Proceedings, Part I, of the 22nd International Conference on Multimedia Modeling* (pp. 562–573). Berlin: Springer.
8. Jin, X., Wu, Y., Li, X., Li, Y., Zhao, G., & Guo, K. (2016). PPViBe: Privacy preserving background extractor via secret sharing in multiple cloud servers. In *8th International Conference on Wireless Communications & Signal Processing, WCSP 2016* (pp. 1–5). Piscataway: IEEE.
9. Jin, X., Yuan, P., Li, X., Song, C., Ge, S., Zhao, G., et al. (2017). Efficient privacy preserving Viola-Jones type object detection via random base image representation. In *Proceedings of IEEE International Conference on Multimedia and Expo (ICME) 2017*. Piscataway: IEEE.
10. Osadchy, M., Pinkas, B., Jarrous, A., & Moskovich, B. (2010). SCiFI – a system for secure face identification. In *31st IEEE Symposium on Security and Privacy, S&P 2010* (pp. 239–254). Piscataway: IEEE.
11. Shashank, J., Kowshik, P., Srinathan, K., & Jawahar, C.V. (2008). Private content based image retrieval. In: *2008 IEEE Computer Society Conference on Computer Vision and Pattern Recognition (CVPR 2008)*. Piscataway: IEEE.
12. Sohn, H., Plataniotis, K. N., & Ro, Y. M. (2010). Privacy-preserving watch list screening in video surveillance system. In *Proceedings of the Advances in Multimedia Information Processing - PCM 2010 – 11th Pacific Rim Conference on Multimedia* (pp. 622–632). Berlin: Springer.
13. Upmanyu, M., Namboodiri, A. M., Srinathan, K., & Jawahar, C. V. (2009). Efficient privacy preserving video surveillance. In *IEEE 12th International Conference on Computer Vision, ICCV 2009* (pp. 1639–1646). Piscataway: IEEE.
14. Viola, P. A., & Jones, M. J. (2001). Robust real-time face detection. In *IEEE 8th International Conference on Computer Vision, ICCV 2011* (p. 747). Piscataway: IEEE.
15. Viola, P. A., & Jones, M. J. (2004). Robust real-time face detection. *International Journal of Computer Vision, 57*(2), 137–154.
16. Wong, W. K., Cheung, D. W. l., Kao, B., & Mamoulis, N. (2009). Secure kNN computation on encrypted databases. In *Proceedings of the 2009 ACM SIGMOD International Conference on Management of Data* (p. 139–152). New York, NY: ACM.
17. Wright, J., Yang, A. Y., Ganesh, A., Sastry, S. S., & Ma, Y. (2009). Robust face recognition via sparse representation. *IEEE Transactions on Pattern Analysis and Machine Intelligence, 31*(2), 210–227.

Near-Duplicate Video Cleansing Method Based on Locality Sensitive Hashing and the Sorted Neighborhood Method

Ou Ye, Zhanli Li, and Yun Zhang

1 Introduction

At present, intelligent video surveillance (IVS) technology has been widely used in various areas in society. In coal mining, for instance, more than 95% of coal mine enterprises in China have installed the intelligent video surveillance system [1]. With the wide utilization of this technology, increasing amounts of dirty video data are emerging, which seriously affect video data quality. The video data quality mentioned here is the data quality of the video data set, which is different from video quality [2]. Video quality focuses on the clarity of video data, but video data quality refers to concept of data quality [3], which emphasizes the extent of video data consistency, correctness, completeness and the minimum being satisfied in the information system. If video data cause the video data set to fail to meet the above criteria of data quality, they are dirty videos in addition to the dirty data [4]. In fact, the dirty data come from a wide range of sources and have various forms; they are generated in the process of data collection, integration, processing, and storage. Near-duplicate video data are one of the most common forms of dirty data, which share the same semantics and their scenes may differ slightly. Wu et al., based on a sample of 24 popular queries, retrieved near-duplicate videos by using the YouTube, Yahoo! Video, and Google Video websites. The result shows that on average there

Please note that the LNICST Editorial assumes that all authors have used the western naming convention, with given names preceding surnames. This determines the structure of the names in the running heads and the author index.

O. Ye (✉) · Z. Li · Y. Zhang
School of Computer Science and Technology, Xi'an University of Science and Technology, Xi'an, Shaanxi, China
e-mail: oye0928@xust.edu.cn

© Springer Nature Switzerland AG 2020
H. Lu, L. Yujie (eds.), *2nd EAI International Conference on Robotic Sensor Networks*, EAI/Springer Innovations in Communication and Computing,
https://doi.org/10.1007/978-3-030-17763-8_12

are 27% redundant videos that are near-duplicates to the most popular version of a video, and for certain queries, the redundancy can be as high as 93% [5]. These near-duplicate videos not only seriously affect the normal use of video, but also affects copyright issues. Therefore, reducing the amount of near-duplicate videos in a video data set so as to ensure the video data quality is an urgent problem that needs to be solved right away.

Recently, most research results have focused on near-duplicate video retrieval, in particular, research into the low-level feature extracting algorithm [6, 7], the video signature algorithm [8], the signature indexing algorithm [9], etc. Although these results can help us to identify the dirty video, they cannot be used to clean the dirty data automatically with data quality criteria. Data cleansing [10, 11] technology is a broadly effective way to improve the data quality, which can be used to clean the dirty data in a data set automatically or change the dirty data to normal data. However, less research has been carried out into the data cleansing issue for videos that have a complex background. In this chapter, the video data set of coal mining is taken as an example. We present a novel near-duplicate video cleansing method based on LSH and SNM to improve the data quality of video data sets with a complex background.

The next section, "Related Works" on near-duplicate video retrieval are introduced; in the third section, a "Near-Duplicate Video Cleansing Method Based on LSH and SNM"; in the fourth section, "Experimentation and Analysis"; finally there is the "Conclusion."

2 Related Work

At present, there are many research results for near-duplicate video retrieval (NDVR). In this aspect, Zobel and Hoad [12] use the color histogram or local feature of video data to determine the similarity between two videos to identify the near-duplicate videos. This type of method is easy to implement and efficiency is high. However, the single low-level feature can easily be interfere with by light, geometric deformables, and other factors. In the aspect of NDVR based on the video signature, extracting the feature of the video from the video-level global signature, the frame-level global signature, the frame-level local signature, and the spatial–temporal signature described in Douze et al. [13] and other literature improves the comprehensiveness and robustness of the low-level feature. However, this type of method involves high computational complexity, and the change of spatial–temporal signature can affect the accuracy. The LSH method can be used to improve the efficiency of the above methods. Liu et al. [14] used LSH to compute the match sequence of the frame between video clips first, and then used the random sample consensus to fit the match sequence, determining the near-duplicate video data. Wang and Liu [15] denoted the similarity of videos by using the ratio of the number of matching shots between videos, and LSH is introduced in a special way to improve the speed of near-duplicate video detection. Liu and Zhu [16] constructed a new hashing structure, and by using an adaptive locality sensitive hashing scheme to index local features in key frames of videos identified the near-duplicate videos. Liu and Zhu [17] used an speeded-up robust feature (SURF) descriptor and two-level matching scheme to generate a relevance score for near-duplicate video detection. In general, the existing methods can be used to identify near-duplicate videos, but they cannot clean them automatically. Therefore, the data quality of the video data set cannot be ensured.

A data cleansing method can be used to improve the data quality. However, these methods mainly focus on the normal data types and cleansing of big data. Because research into NDVR started relatively later, and it pays little attention to the data quality of video data sets that have a complex background, there are fewer research results for video cleansing, and it is difficult to ensure the video data quality.

In this chapter, a near-duplicate video cleansing method based on LSH and SNM is presented to address the above problem and to improve video data quality with a complex background. In this method, the SURF descriptor is extracted from the video to represent the video feature first of all, and then the sorted candidate set is built by using the LSH function. On this basis, the near-duplicate videos are cleaned automatically by using SNM.

3 Near-Duplicate Video Cleansing Method Based on LSH and SNM

Suppose that we have a video data set $D = \{V_1, V_2, V_3, \ldots, V_n\}, \forall V_i \in D, V_i \neq V_j$, where V_i and V_j denote any ith and jth video data in D. If the similarity between V_i and V_j is less than threshold δ, they are near-duplicate videos and dirty videos.

On this basis, to ensure video data quality, near-duplicate videos must be cleaned, and the video data set should satisfy the criteria of consistency, correctness, completeness, and the minimality, especially the minimality. In other words, the aim is to have no near-duplicate videos in the video data set by cleaning the dirty videos, and to ensure the completeness of the data set. Therefore, the objective function of near-duplicate video cleansing can be described as (1):

$$f_{\mathrm{ndv}}(D) = \begin{cases} D - \{V_j\}, & \text{if } V_i \approx V_j, V_i \in D, V_j \in D \\ D, & \text{otherwise} \end{cases} \quad (1)$$

According to Eq. (1), we can find that near-duplicate videos first need to be identified; in addition, the representative video V_i needs to be preserved, and the near-duplicate video V_j must be merged and deleted from video data set D.

3.1 Key Frame Extraction of Video Data

Because video data consist of frames, and the scale of frames is normally large, if we measure the similarity between two videos by each frame, the computational complexity is high. To address this issue, we can represent the low semantics of the video by extracting the key frames from the video data.

Suppose that any video V_i in D contains several frames f, the video can be denoted as $V_i = \{f_1, f_2, \ldots, f_m\}$. Among those, any two frames are ordered. On this basis, for every video, we can extract the first frame, an average frame, and the last frame as the key frames. The average frame can be extracted from the video by using Eq. (2).

$$f_{\text{ave}} = \left\lceil \left(\sum_{j=0}^{m-1} f_j \right) / m \right\rceil \tag{2}$$

where m denotes the number of frames in each video, and f_j denotes the pixel matrix for the jth frame.

As f_{ave} may not be contained in the video, we can select one frame that is closest to f_{ave} as the average key frame by using Eq. (3).

$$fk = \underset{j=0,1,2,\ldots,m-1}{\arg\min} \left(|fj - f_{\text{ave}}| \right) \tag{3}$$

Finally, the key frame set of the video can be obtained by using (4).

$$\underset{i \in [0,n-1]}{\text{KFrame}} (V_i) = \{f_0, f_k, f_{m-1}\} \tag{4}$$

3.2 Feature Extraction of Local Key-Points

The video-level and frame-level global signatures are difficult to extract to identify near-duplicate video data, because of light, geometric deformables, and other factors in video data that have a complex background. In this chapter, local key-points featured at the frame level are used to identify weather videos and are near-duplicate, because it can describe features in higher detail than the global signature.

Nowadays, scale invariant features transform, SURF, and Harris/Forstner are the main local key-point descriptors, which can be used to identify the near-duplicate video data. Because the SURF local key-points descriptor has the scale-invariant characteristic, and it can combine the integral image and Hessian matrix, its efficiency is high. In this chapter, the SURF [18] is used to represent the feature of video data.

To extract the SURF feature from videos that have a complex background, the integral image is first calculated by Eq. (5). In this image, the calculation of the sum of pixels in any rectangular region is simple.

$$I_{\sum}(X) = \sum_{i=0}^{i \leq x} \sum_{j=0}^{j \leq y} I(i, j) \tag{5}$$

For any point $X = (x, y)$ on the integral image, the definition of δ scale is shown in Eq. (6), where $L_{xx}(X, \delta)$ is the convolutional result of the second derivative of Gaussian filters and $I(x, y)$.

$$H(X, \delta) = \begin{bmatrix} L_{xx}(X, \delta) & L_{xy}(X, \delta) \\ L_{xy}(X, \delta) & L_{yy}(X, \delta) \end{bmatrix} \tag{6}$$

Because the Gaussian filter of discretization may be out of shape, box-filtering can be close to the Gaussian filter to be calculated. Suppose that the convolutions of the three directions are D_{xx}, D_{yy}, and D_{xy}, the estimates of the Hessian matrix of each point on the scale space can be calculated by using Eq. (7).

$$\text{Det}(\Delta H) = D_{xx}D_{yy} - \left(0.9D_{xy}\right)^2 \tag{7}$$

If the value of $\text{Det}(\Delta H)$ is positive, and the value of the feature is also positive, this point is the local key point.

Because each video contains several key frames, and each key frame contains several key points, we need to use the PCA algorithm to extract the principal component of key points at the video level. Suppose the SURF matrix of ith key frame is described as $\boldsymbol{Dp_i}$, and the feature matrix of the video can be viewed as the training sample set, which is made up of several $\boldsymbol{Dp}$. To extract the principal component from the training sample set to represent the feature of the video, the average vector of the sample set needs to be calculated by using Eq. (8).

$$\overline{\boldsymbol{Dp}} = \frac{1}{n}\sum_{i=0}^{n-1}\boldsymbol{Dp_i} \tag{8}$$

The covariance matrix of sample set $\text{Cov}(\boldsymbol{Dp_i})$ can be calculated by Eq. (9).

$$\text{Cov}\left(\boldsymbol{Dp_i}\right) = \frac{1}{n}\sum_{i=0}^{n-1}\left(\boldsymbol{Dp_i} - \overline{\boldsymbol{DP}}\right)\left(\boldsymbol{Dp_i} - \overline{\boldsymbol{Dp}}\right)^T \tag{9}$$

Through the eigenvalue decomposition and eigenvalue sorting for the covariance matrix of sample set $\text{Cov}(\boldsymbol{Dp_i})$, the eigenvector of the video that corresponds to the top k eigenvalue ($k = 3$ in this chapter) can be obtained by using Eq. (10).

$$\text{Cov}\left(\boldsymbol{Dp_i}\right)\boldsymbol{W} = \lambda\boldsymbol{W} \tag{10}$$

We label the eigenvector $V_{\text{LKP}(i)}$, and $V_{\text{LKP}(i)}$ is described in Eq. (11).

$$V_{\text{LKP}(i)} = \left(w_{l(0)}, w_{l(1)}, \ldots, w_{l(k-1)}\right) \tag{11}$$

where w_l denotes the lth eigenvalue of ith video data ($l = 3$ in this chapter).

3.3 Near-Duplicate Video Cleansing Method Based on LSH and SNM

To index each video in the video data set rapidly and clean the near-duplicate video, the LSH method and SNM are combined to clean dirty video automatically.

First, we use the LSH function to index near-duplicate videos that have a similar eigenvector. The hash function is shown as Eq. (12).

$$h\left(V_{\mathrm{LKP}(i)}\right) = ha, b\left(V_{\mathrm{LKP}(i)}\right) = \left\lfloor \frac{\mathbf{a} \bullet V_{\mathrm{LKP}(i)} + \mathbf{b}}{\mathbf{W}} \right\rfloor \tag{12}$$

where random vector $\boldsymbol{a}$ follows normal distribution, and random vector $\boldsymbol{b}$ follows [0, $\boldsymbol{w}$] uniform distribution ($\boldsymbol{w} = 6$ in this chapter). Among these, $\boldsymbol{w}$ determines the size of the hash bucket.

Then, the sorted candidate set $S(V_{\mathrm{LKP}(i)}) = \{V_{\mathrm{LKP}(1)}, \ldots, V_{\mathrm{LKP}(L)}\}$ of the ith video Vi can be obtained by using the Hamming distance and $h(V_{\mathrm{LKP}(i)})$.

On this basis, the idea of SNM can be used to clean near-duplicate video from the candidate set. In the $S(V_{\mathrm{LKP}(i)})$, if the distance between the neighborhood hash values that correspond to the eigenvectors of Vi and Vj is less than the threshold, Vi and Vj are merged, and Vj is deleted. The sorted neighborhood near-duplicate video cleansing method is shown below.

```
Input: S (V_LKP(i) )

Output: f_ndv (D)

Auxiliary Variables: row number i, threshold value
delta_threshold, Window Size Wnd_Size.

Initialization: D, i=0,
delta_threshold=0.0016, Wnd_Size=1, D'={ }

Begin

    While i<len(S(V_LKP(i))):

        if(i+Wnd_Size)<len(S(V_LKP(i))):

//video signature vectors of neighborhood video in window

            FirstData = [i]

            SecData = [i+Wnd_Size]

            dis = abs(|SecData-FirstData|)

            //According to equation (1) and (2)

            if dis < delta_threshold:

FirstData → V_i

SecData → V_{i+W-nd_Size}
```

```
//combining two near-duplicate videos
```

$V_i\ =\ V_i\ \cup\ V_{i\ +\ W\text{-}nd\_Size}$

```
                    Update S(V_LKP(i))
                    Update D
```

$f_{ndv}(D)\ =\ D$

```
                i=i + 1

            else:

                i=i + 1

                continue

        else:

            break

    return f_ndv(D)
```

End

4 Experimentation and Analysis

To verify the viability of the method that is mentioned in this chapter, precision and recall indicators are used to show the performance of the method by Eqs. (13) and (14).

$$\text{Precision} = T/P = T/(T + F) \tag{13}$$

$$\text{Recall} = T/R \tag{14}$$

where T denotes the correct results, F denotes the false results, and R denotes the real correct results.

In this section, we select 73 near-duplicate videos from coal mining as the test data, which have more noise and uneven light; thus, the background is complex. By contrasting with the method in Wang and Liu [15] to compare the video cleansing results of the global feature (GIST descriptor in [15]) and the local feature (SURF descriptor in this chapter), all methods are implemented by using the Python language. The video cleansing results are shown in Figs. 1, 2, and 3.

Because the SURF of the video can be described in higher detail, the cleansing results and recall of SURF are better than those of the GIST, as shown in Figs. 1 and 3. Although the amount of video data is small, the global feature can be used to identify near-duplicate video. However, with the amount of videos increasing, noise, light and other factors affect the precision of video cleansing; thus, the precision of GIST is less than that of SURF (Fig. 2). Because local key-points have great computational complexity, the duration of SURF is longer than that of GIST (Fig. 4).

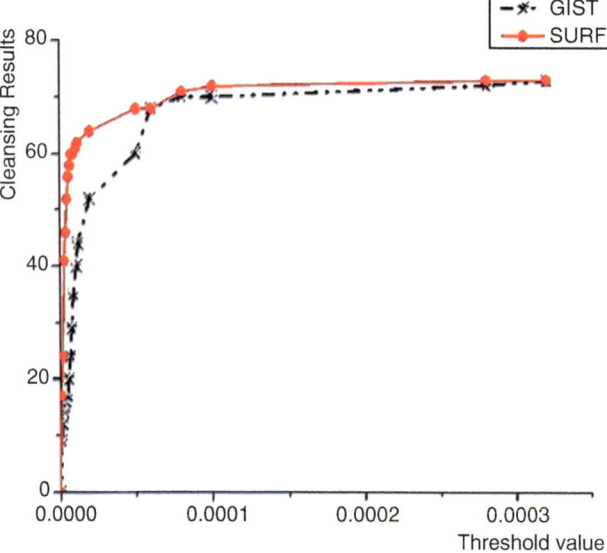

Fig. 1 Cleansing results of the two methods

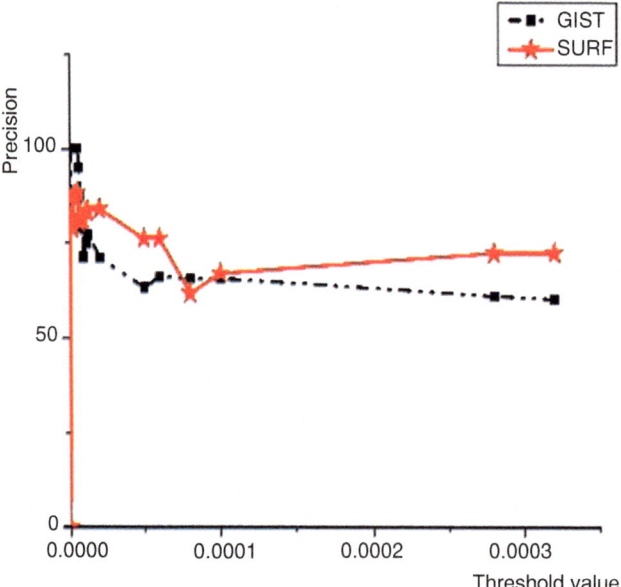

Fig. 2 The precision results of the two methods

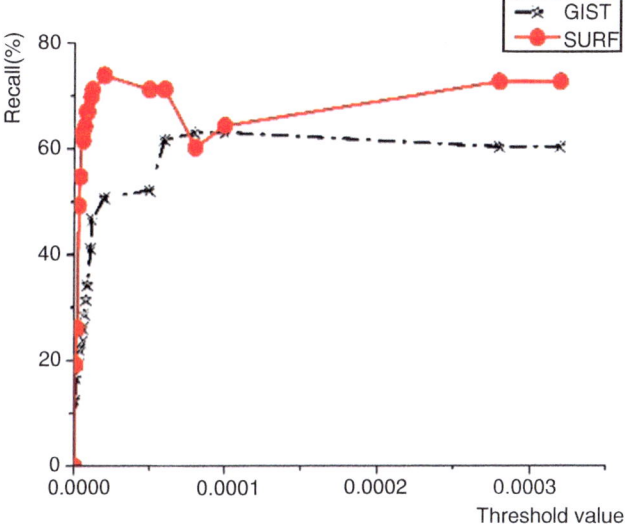

Fig. 3 The precision results of the two methods

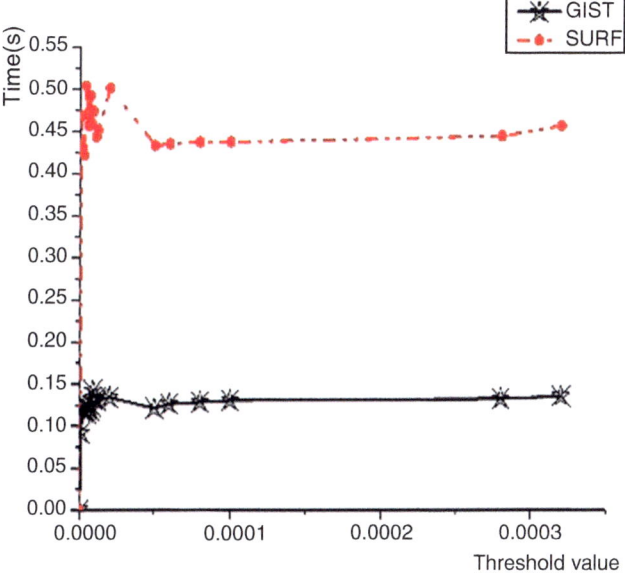

Fig. 4 Execution time of the two methods

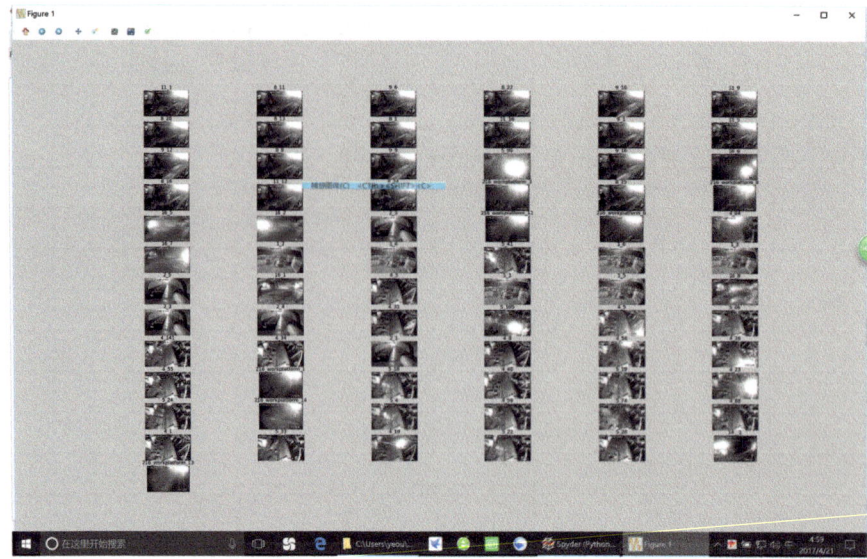

Fig. 5 The cleaning result

Finally, the experiment result of the near-duplicate video cleansing method based on LSH and SNM is shown in Fig. 5. From the result, we can find that, even though light and noise exist, near-duplicate videos can be cleaned using the presented method, and it can improve video data quality.

5 Conclusion

In this chapter, a near-duplicate video cleansing method based on LSH and SNM is presented, which can be used to improve the data quality of a video data set. In this method, the SURF descriptor is first extracted from the video to represent the video feature, and then the sorted candidate set is built by using LSH. On this basis, near-duplicate videos are cleaned automatically by using SNM. Finally, the simulation experiments are implemented to show that the method presented in this chapter is effective, and can be used to clean near-duplicate videos automatically and improve video data quality. In the future, we will use a deep learning model to extract the feature of the video to improve the accuracy of video cleansing and further ensure the data quality of the video.

Acknowledgements This work was supported in part by the Shannxi Provincial Department of Education special scientific research project (No.16JK1505).

References

1. Wang, W., & Zhang, L. (2013). Application and research of security data mining techniques in coal mine mobile video monitoring system (in Chinese). *Coal Technology, 9*, 101–103.
2. Chikkerur, S., Sundaram, V., Reisslein, M., et al. (2011). Objective video quality assessment methods: A classification, review, and performance comparison. *IEEE Transactions on Broadcasting, 57*(2), 165–182.
3. Ringler, A. T., Hagerty, M. T., Holland, J., et al. (2015). The data quality analyzer: A quality control program for seismic data. *Computers & Geosciences, 76*, 96–111.
4. Kim, W., Choi, B. J., Hong, E. K., et al. (2003). A taxonomy of dirty data. *Data Mining and Knowledge Discovery, 7*(1), 81–99.
5. Wu, X., Ngo, C., Hauptmann, A., et al. (2009). Real-time near-duplicate elimination for web video search with content and context. *IEEE Transactions on Multimedia, 11*(2), 196–207.
6. Huang, Z., Shen, H. T., Shao, J., Zhou, X., & Cui, B. (2009). Bounded coordinate system indexing for real time video clip search. *ACM Transactions on Information Systems, 27*(3), 17–33.
7. Huang, Z., Hu, B., Cheng, H., Shen, H. T., Liu, H., & Zhou, X. (2010). Mining near-duplicate graph for cluster-based reranking of web video search results. *ACM Transactions on Information Systems, 28*(4), 22.
8. Zhou, X., Chen, L., Bouguettaya, A., Xiao, N., & Taylor, J. A. (2009). An efficient near duplicate video shot detection method using shot-based interest points. *IEEE Transactions on Multimedia, 11*(5), 879–891.
9. Liu, J., Huang, Z., Cai, H., et al. (2013). Near-duplicate video retrieval: Current research and future trends. *ACM Computing Surveys, 45*(4), 44–46.
10. Rahm, E., & Do, H. H. (2000). Data cleaning: Problems and current approach. *IEEE Data Engineering Bulletin, 23*(4), 3–13.
11. Minnich, A., Abu-El-Rub, N., Gokhale, M., et al. (2016). Clear view: Data cleaning for online review mining. In *IEEE/ACM International Conference on Advances in Social Networks Analysis and Mining (ASONAM)* (pp. 555–558). San Francisco: IEEE Press.
12. Zobel, J., & Hoad, T. C. (2006). Detection of video sequences using compact signatures. *ACM Transactions on Information Systems, 24*(1), 1–50.
13. Douze, M., Jégou, H., & Schmid, C. (2010). An image-based approach to video copy detection with spatiotemporal post-filtering. *IEEE Transactions on Multimedia, 12*(4), 257–266.
14. Liu, S., Zhu, M., & Zheng, Q. (2010). A detection method for near duplicate video clips based on content similarity (in Chinese). *Journal of University of Science and Technology of China, 40*(11), 1130–1135.
15. Wang, H., & Liu, X. (2012). Near-duplicate web video detection based on locality sensitive hashing (in Chinese). *Application Research of Computers, 29*(5), 1954–1958.
16. Liu, D., & Zhu, M. (2013). A fast algorithm for near-duplicate video detection (in Chinese). *Journal of Chinese Computer Systems, 34*(6), 1400–1406.
17. Liu, D., & Zhu, M. (2015). A computationally efficient algorithm for large scale near-duplicate video detection. In *International Conference on Multimedia Modeling (MMM 2015)* (pp. 481–490). Basel: Springer.
18. Bay, H., Tuytelaars, T., & Van Gool, L. (2008). Speeded-up robust features (SURF). *Computer Vision and Image Understanding, 110*(3), 346–359.

A Double Auction VM Migration Approach

Jinjin Wang, Yonglong Zhang, Junwu Zhu, and Yi Jiang

1 Introduction

With the rapid development of information and data in the Internet age, computational domain including science, engineering, and business need to process large-scale, massive data. In this case, the concept of cloud computing is proposed. Cloud computing is a further development of distributed computing, parallel processing, and grid computing [13]. It is a system based on Internet computing that can provide hardware services, infrastructure services, platform services, software services, and storage services to a variety of Internet applications [6, 7].

Virtualization is a key technology of cloud computing, which can turn a host into multiple virtual hosts which have different computer systems resources that can support applications. Virtualization has lot of advantages. By reducing the number of physical hosts by turning the physical host into a virtual host, energy consumption is reduced and energy efficiency is achieved. In addition, virtualization is a cost-effective technology [3]. However, load balancing is a challenge. There are overloaded hosts and underutilized hosts. When too many VMs are running on one host, the host will become overloaded and cause exceptions. VM migration can solve this problem. In this case, we must select one or more VMs to migrate once the host become overloaded and find the best destination host for these selected VMs which to be migrated. Double auction is widely used in the field of artificial intelligence to solve the problem of resource competition [5, 14]. In this paper,

J. Wang · Y. Zhang · Y. Jiang
College of Information Engineering, Yangzhou University, Yangzhou, Jiangsu, China

J. Zhu (✉)
College of Information Engineering, Yangzhou University, Yangzhou, Jiangsu, China

Department of Computer Science, University of Guelph, Guelph, ON, Canada
e-mail: jwzhu@yzu.edu.cn

H. Lu, L. Yujie (eds.), *2nd EAI International Conference on Robotic Sensor Networks*, EAI/Springer Innovations in Communication and Computing,
https://doi.org/10.1007/978-3-030-17763-8_13

we proposed a VM migration algorithm based on double auction, which considers communication cost.

The rest of this paper is organized below. Section 2 reviews some related works. In Sect. 3, we detail system model. We introduce VMs-GSA and VMM-DAM in Sect. 4. Section 5 is simulation results and conclusion.

2   Related Work

In recent years, VM migration in the data center has been widely studied and has become a hot topic. The process of VM migration mainly involves energy consumption and communication costs. In order to improve energy efficiency of the data center, Tao et al. [10] proposed a new algorithm named BGM-BLA for VM migration, which considered three factors including energy consumption, communication cost, and migration cost. The algorithm in [10] targets these three factors has two parts. The first part generates bucket codes. The second part is to learn and mutate to reduce communication cost and migration cost based on the original bucket codes and output the Pareto set of solutions. Beloglazov et al. [2] proposed energy-aware allocation heuristics which provides the resources of servers or hosts to client applications while guaranteeing quality of service (QoS). In [8], the authors proposed three VM migration schemes which take the traffic factor and VM clustering into account, which are the improvement of papers [12]. VM clustering includes two steps. First step forms the VM graph on same host, in which vertexes are VMs and edges are communications between VMs. Then, form clusters. In addition, many studies have considered communication factors in the migration of VMs, such as [4] and [11].

Zhang et al. [15] apply the genetic algorithm (GA) and artificial bee colony (ABC) to the problem of VM migration problem and aim to find an approximate optimal solution through repeated iterations. The GA first generated a population of PopSize chromosomes which is a vector represented the mapping of VMs and servers. Then, cross and mutate the chromosomes and compare with the previous chromosome to select smaller chromosomes with smaller fitness value. In [1], migration algorithm is investigated, which is a new system based on matching game theory. The paper applied firefly algorithm to energy-aware VM migration, which migrates the maximally loaded VM to the least loaded active node.

3   System Model

We consider that there are K hosts in data center and the relation between these hosts can be denoted as an undirected graph $DC = (H, D)$, where $H = \{h_1, h_{2,...,}h_k\}$ represents the set of hosts and $\forall(h_i, h_j) \in D$ represents the edge between hosts h_i and h_j. The weight $W(h_i, h_j)$ is the communication distance.

Each host $h_j \in H$ can be represented by a 3-tuple $(\tau(h_j), sl(h_j), V(h_j))$, where $\tau(h_j)$ is the threshold of h_j, i.e., if h_j's load exceeds this value $\tau(h_j)$, h_j is overloaded. $sl(h_j)$ indicates the safety limit of h_j, i.e., if h_j's load is lower than this value $sl(h_j)$, h_j is underutilized. $V(h_j) = \{v_1^j, v_2^j, \ldots, v_n^j\}$ is the set of VMs residing on h_j. The communication relation of VMs residing on h_j is represented by an undirected graph $G = (V, E)$, where $V = V(h_j)$ and $E = \{(v_i^j, v_l^j) | \text{where } v_i^j \text{ and } v_l^j \text{ exist communications}\}$ is the set of edges. The weight $W'(v_i^j, v_l^j)$ represents VM traffic. $\forall v_l^j \in V(h_j)$ that communicate with v_i^j are called v_i^j's neighbor, i.e., $N(v_i^j) = \{v_l^j | (v_i^j, v_l^j) \in E, \forall v_l^j \in V(h_j)\}$.

$v_i^j \in V(h_j)$ have two attributes denoted by a 2-tuple $(O(v_i^j), C(v_i^j))$, where $O(v_i^j)$ denotes the size of occupied resources of v_i^j. $C(v_i^j) = \sum_{v_l^j \in N(v_i^j)} W'(v_i^j, v_l^j)$ is the traffic of v_i^j. The residing set of overloaded hosts is represented by $H^+ = \{h_j | \sum_{i \in V(h_j)} O(i) > \tau(h_j), h_j \in H\}$ and $H^- = \{h_j | \sum_{i \in V(h_j)} O(i) \leq sl(h_j), h_j \in H\}$ denotes the set of underutilized hosts. We use $h_j^+ \in H^+$ to denote overloaded host h_j and $h_j^- \in H^-$ to denote underutilized host h_j.

Our work is mainly focused on the communication costs of VMs migration in data center. Let mapping function $\sigma : VM \rightarrow H^-$ denotes the mapping between VMs and underutilized hosts. Let vm_i denote the selected VM from overloaded host. The communication cost that vm_i matches $h_j^- \in H^-$ is defined as below

$$Cost_i = \sum_{v_l \in N(vm_i)} W'(vm_i, v_l) W(h_{\sigma(i)}^-, h_{\sigma(l)}^-) \qquad (1)$$

4 VM Migration Algorithm

In this section, VM migration algorithm based on heuristic consists of two parts: (1) selecting VMs from overloaded hosts to migrate and we proposed VMs-GSA to determine these VMs, (2) obtaining the mapping between VMs and underutilized hosts and we employed VMM-DAM to obtain it.

4.1 VMs-GSA Design

The idea of VMs-GSA is to select the VMs resided on $h_j^+ \in H^+$ with smaller traffic and smaller occupied resource to migrate, thereby reducing communication costs. Therefore, the algorithm will be executed as follows. First of all, the VMs residing on h_j^+ are sorted by $O(v_i^j)C(v_i^j)$ in the ascending order. Then, starting from v_i^j with the highest $O(v_i^j)C(v_i^j)$ value, select the VMs in sequence and put them in the list W until the host is not overloaded.

Algorithm 1 VMs-GSA (one host h_j^+)

Input: The set of VMs on h_j^+, $V(h_j)$
Output: The set of VMs selected from h_j^+, W
1: Sort VMs by $O(v_i^j)C(v_i^j)$ in ascending order
2: **for** $u = 1$ to $|V(h_j)|$ **do**
3:     Assume the u-th total communication is $O(v_u^j)C(v_u^j)$
4:     **if** h_j^+ is overloaded **then**
5:         $W = W \cup \{v_u^j\}$
6:     **end if**
7: **end for**
8: **return** W

4.2 VMM-DAM Design

Auction Model We consider that the allocation process of finding the destination hosts for VMs which to be migrated is modeled as an auction process. Auction market consists of three entities. The buyers refer to the VMs which to be migrated. The sellers are the underutilized hosts. The third-party auctioneer mainly solves the mapping problem and final payment. Buyer $vm_i \in VM$ submits a bid represented by B_i to auctioneer and B_i can be denoted by a 2-tuple:$(O(vm_i), v_i)$, where $O(vm_i)$ is the size of vm_i's resource demand and v_i indicates vm_i's valuation. $B = \{B_1, B_2, \ldots, B_N\}$ denotes the bids of all buyers. Seller $h_j^- \in H^-$ submits a bid S_j to the auctioneer. S_j is denoted as a 2-tuple: $(o_j, p_j(m_j))$, where $p_j(m_j)$ denotes the unit price of the resource o_j provided by h_j^-, which is piecewise constant function. m_j is the quantity sold of h_j^-. The bids of all sellers are denoted as $S = \{S_1, S_2, \ldots, S_M\}$.

Allocation Algorithm Design Let $d_i = v_i/\sqrt{O(vm_i)}$ be vm_i's bid density. Firstly, we sort the VMs according to d_i in the descending order. Let L be the sorted VMs list. Then, select VM from the sorted list L in turn to match seller. Suppose the selected VM currently is vm_i. The host h_{j*}^- which satisfies two conditions of vm_i with maximum revenue increment $RI_{ij} = v_i - O(vm_i)p_j(m_j)$ is matched with vm_i. One condition is that the host can meet demand of vm_i and another is that the ask price of the host is more than vm_i's valuation. Next, the algorithm finds other VMs that reside on the vm_i's original host from L and put them in list $L_i' \subset L$. Then, select VM from L_i' in turn to match seller. Suppose the selected VM currently is vm_u. If h_{j*}^- satisfies the conditions of vm_u, vm_u is matched with h_{j*}^-. Otherwise, the algorithm finds other hosts which satisfy conditions of vm_u. The host with maximum $\varphi_{uk} = \alpha RI_{uk} - \beta Cost_u$ becomes the destination host of vm_u, where α and β are the weight coefficients. The algorithm loops until all the VMs in L are matched.

Algorithm 2 VMs allocation algorithm

Input: The set of buyers' bids, B; The set of sellers' bids, S
Output: The matrix which is the mappings result of buyers and sellers, X
1: Sort buyers by d_i in descending order and put the sorted buyers in L
2: **for** $vm_i \in L$ **do**
3:    **if** exist hosts which satisfy two conditions of vm_i **then**
4:       Choose j with the greatest RI_{ij} to matched vm_i and $x_{ij} = 1$
5:       Find other VMs that reside on the vm_i's original host from L and put them in list L'_i,
6:       **for** $vm_u \in L'_i$ **do**
7:          **if** h_j^- satisfy two conditions of vm_u **then**
8:             h_j^- is matched with vm_u and $x_{uj} = 1$
9:          **else**
10:             **if** exist hosts which satisfy two conditions of vm_u **then**
11:                Choose k with the greatest φ_{uk} to matched vm_u and $x_{uk} = 1$
12:             **else**
13:                vm_u failed to match host
14:             **end if**
15:          **end if**
16:       **end for**
17:    **else**
18:       vm_i failed to match host
19:    **end if**
20:    $L = L \backslash L'_i$
21: **end for**
22: **return** X

Scheme of Payment We employ "Vickery" price to the payment of buyers [9]. The "Vickery" price of vm_i is defined as the value of the size of vm_i's demand multiplied by the highest bid density of vm_l among losers who would become the winner if vm_i would not participate in the auction. The winner vm_i' payment c_i to its matched seller is $c_i = \max\{d_l\sqrt{O(vm_i)}, O(vm_i)p_j(m_j)\}$. The payment b_j to the winner h_j^- is the sum of payments of all the VMs that matched h_j^-.

5 Results and Conclusion

In this section, we simulated VMs migration in the data center and summarized the paper. Simulation experiment has shown that compared with the random algorithm, the total amount of communication generated by VMs-GSA reduces about 35% and the VMs-GSA is approximately similar to the enumeration algorithm. However, the computational complexity of the enumeration algorithm is very huge. The load of overloaded hosts drops significantly to the threshold after the migration by VMM-DAM and VM migration is almost 100% successful and the communication generated by VMM-DAM is small.

We investigated traffic-aware VM migration problem in data center. The VM migration algorithm consists of two parts. We designed VMs-GSA in the first part to select VMs with the low communication costs, which greatly reduces the communication generated by VM migration. VMM-DAM is applied to the second part of VM migration to match hosts with the low communication costs. Simulation experiment has shown that the VMs-GSA and VMM-DAM are traffic-aware and effective.

Acknowledgements This work was supported by the National Nature Science Foundation of China under Grant 61170201, Grant 61070133, and Grant 61472344, in part by the Innovation Foundation for graduate students of Jiangsu Province under Grant CXLX12 0916, in part by the Natural Science Foundation of the Jiangsu Higher Education Institutions under Grant 14KJB520041, in part by the Advanced Joint Research Project of Technology Department of Jiangsu Province under Grant BY2015061-06 and Grant BY2015061-08, and in part by the Yangzhou Science and Technology under Grant YZ2017288 and Yangzhou University Jiangdu High-end Equipment Engineering Technology Research Institute Open Project under Grant YDJD201707.

References

1. Azougaghe, A., Oualhaj, O. A., & Hedabou, M. (2017). Many-to-one matching game towards secure virtual machines migration in cloud computing. In *International Conference on Advanced Communication Systems and Information Security* (pp. 1–7). Piscataway: IEEE.
2. Beloglazov, A., Abawajy, J., & Buyya, R. (2012). Energy-aware resource allocation heuristics for efficient management of data centers for cloud computing. *Future Generation Computer Systems, 28*(5), 755–768.
3. Goldberg, R. P. (1974). Survey of virtual machine research. *Computer, 7*(6), 34–45.
4. Kansal, N. J., & Chana, I. (2016). Energy-aware virtual machine migration for cloud computing—a firefly optimization approach. *Journal of Grid Computing, 14*(2), 327–345.
5. Lu, H., Li, B., Zhu, J., & Li, Y. (2017). Wound intensity correction and segmentation with convolutional neural networks. *Concurrency and Computation Practice and Experience, 29*(6), e3927.
6. Lu, H., Li, Y., Chen, M., Kim, H., & Serikawa, S. (2018). Brain intelligence: Go beyond artificial intelligence. *Mobile Networks and Applications, 23*(2), 368–375.
7. Lu, H., Li, Y., & Mu, S. (2018). Motor anomaly detection for unmanned aerial vehicles using reinforcement learning. *IEEE Internet of Things Journal, 5*(4), 2315–2322.
8. Reguri, V. R., Kogatam, S., & Moh, M. (2016). Energy efficient traffic-aware virtual machine migration in green cloud data centers. In *IEEE International Conference on Big Data Security on Cloud* (pp. 268–273). Piscataway: IEEE.
9. Sun, Z., & Zhu, Z. (2015). A combinatorial double auction mechanism for cloud resource group-buying. In *2014 IEEE 33rd International Performance Computing and Communications Conference (IPCCC)* (pp. 1–8). Piscataway: IEEE.
10. Tao, F., Li, C., & Liao, T. W. (2016). BGM-BLA: A new algorithm for dynamic migration of virtual machines in cloud computing. *IEEE Transactions on Services Computing, 9*(6), 910–925.
11. Tso, F. P., Hamilton, G., Oikonomou, K., & Pezaros, D. P. (2013). Implementing scalable, network-aware virtual machine migration for cloud data centers. In *IEEE Sixth International Conference on Cloud Computing* (pp. 557–564). Piscataway: IEEE.

12. Vu, H., & Hwang, S. (2014). A traffic and power-aware algorithm for virtual machine placement in cloud data center. *International Journal of Grid and Distributed Computing, 7*(1), 21–32.
13. Wang, L., Laszewski, G. V., & Younge, A. (2010). Cloud computing: A perspective study. *New Generation Computing, 28*(2), 137–146.
14. Xu, X., He, L., Lu, H., Gao, L., & Ji, Y. (2018). Deep adversarial metric learning for cross-modal retrieval. *World Wide Web-Internet and Web Information Systems, 22*(2), 657–672.
15. Zhang, W., Han, S., He, H., & Chen, H. (2017). Network-aware virtual machine migration in an overcommitted cloud. *Future Generation Computer Systems, 76*, 428–442.

An Auction-Based Task Allocation Algorithm in Heterogeneous Multi-Robot System

Jieke Shi, Zhou Yang, and Junwu Zhu

1 Introduction

With the fast development of technologies, robots are applied to more and more field. However, in terms of current robotics development, the single-robot system has limitations in getting information and solving problems. What's more, nowadays the environments faced by the robots are always dynamic, real-time, complex adversarial, and stochastic. Facing such complex tasks and the changing working environment, it is hard for a single-robot system to perform operations.

Compared to a single-robot system, a multi-robot system has the features about the distribution of time, space, information, and resources. In order for multiple robots to cooperate efficiently, the key problem we need to research on is how to allocate tasks properly according to working environment and situations. This is a fundamental problem in robots system researchers, also called multi-robot task allocation, MRTA.

When the number and states of the robots are clear, MRTA problem can be thought of as an assignment problem if the number and states of tasks are also clear. Usually, an accomplished task can provide some payoff to the system. In some special working environment, such as the robot rescue tasks, accomplishing a task will not provide the system with any positive payoff. But if the task is always delayed, it will cause damage to the system continuously.

J. Shi · Z. Yang
College of Information Engineering, Yangzhou University, Yangzhou, Jiangsu, China

J. Zhu (✉)
College of Information Engineering, Yangzhou University, Yangzhou, Jiangsu, China

Department of Computer Science, University of Guelph, Guelph, ON, Canada
e-mail: jwzhu@yzu.edu.cn

© Springer Nature Switzerland AG 2020   149
H. Lu, L. Yujie (eds.), *2nd EAI International Conference on Robotic Sensor Networks*, EAI/Springer Innovations in Communication and Computing,
https://doi.org/10.1007/978-3-030-17763-8_14

In this paper, we focus on task allocation problem in rescuing environment. So our aim is to minimize the whole damage to the system caused by sudden catastrophic events.

In dynamic, changing, and stochastic working environments, a successful task scheduling strategy fulfills the following two points:

- *Optimization*: Finding a nearly optimized task allocation scheme within a certain time constraint.
- *Dynamic*: Ensuring the efficiency during the whole working time. In other words, transferring the allocation outcome properly according to the changing of the working environment.

In this paper, in response to the above two requirements, based on auction, we propose a merging method between centralized and distributed methods.

We make three contributions in this work. Firstly, we propose a dynamic auction method for differentiated tasks under cost rigidities, DAMCR. When the number of tasks waiting to be assigned is up to some certain threshold value, DAMCR is activated. Secondly, we prove that DAMCR is ϵ-optimal. Finally, we design detailed experiments and demonstrate the results to analyze our method.

2 Related Work

In the field of multi-robot system, we can use the dual programming representation process for solving decision problems [1]. The traditional dual programming algorithm is generally based on the global static environment known and determined, but the results of task allocation cannot necessarily achieve the goals in the ideal time [5]. In order to improve the efficiency of search and rescue (SAR) in disaster relief, some researchers used auction-based task allocation scheme to develop a cooperative rescue plan [2, 10].

As to dealing with the problem of robot collaboration in the distributed environment, auction algorithms are a feasible method for task assignment that have been shown to efficiently produce suboptimal solutions [4]. The traditional way of computing the winner is to have a central system acting as the auctioneer to receive and evaluate each bid in the fleet. Once all of the bids have been collected, a winner is selected based on a predefined scoring metric [9]. Considering multiple resources of the robots and limited robot communication range, Lu [8] applied the idea of second-price auction to determine the final price and the number of provisioned VMs in the double auction.

The downside of these approaches is that the bids from each agent must somehow be transmitted to the auctioneer [7, 11]. In the rescue robot system there exist a number of rescue robots, the robots need to complete the corresponding task to ensure the system loss minimum. In order to obtain an ideal task allocation scheme, we must take full account of the benefits and costs of completing the task [3]. A common method to avoid communication limitation is to sacrifice mission performance by running the auction solely within the set of direct neighbors of the auctioneer [6].

3   Model Description and Notations

3.1   *Definitions of Notations and Model*

The dynamics of the environment are mainly reflected in the changes in some factors of the work environment as time progresses. We define E_t to represent the multi-robot system work environment at time t. Mathematically speaking, E_t can be formalized as triples:

$$E_t = <R, T_1, T_2> \tag{1}$$

R is the set of all the robots in this system, it can be formalized as below:

$$R = <r_1, r_2, \ldots, r_N> \tag{2}$$

r_i is the ith robot in the system. $N = |R|$ represents the total number of robots. A robot r_i can be described as below:

$$r_i = <robID_i, state_i, place_i, t> \tag{3}$$

In this formula, $robID_i$ uniquely identifies a robot. $place_i$ represents the robot's location. $state_i$ indicates the status of the robot r_i, whose value rules are demonstrated below:

$$state_i = \begin{cases} 0, & \text{if the robot is free;} \\ taskID_j, & \text{if } task_j \text{ is assigned to it.} \end{cases} \tag{4}$$

In formula (1), T_1 is the set of tasks that have been allocated while T_2 is the set of the tasks waiting to be assigned. Apparently, the set T consisting of all the tasks in the system is $T = T_1 \cup T_2$. An element in those sets, in other words, a task

$$task_j = <taskID_j, state_j, place_j, C_j(AR_j^t), t> \tag{5}$$

$taskID_j$ uniquely identifies a task. $place_j$ represents the robots location. Similar to a Cartesian coordinate system, we use (x_j, y_j) to represent $place_i$. $state_i$ indicates the status of the robot t_j, whose value rules are demonstrated below:

$$state_j = \begin{cases} 0, & \text{if the task is not assigned;} \\ taskID_j, & \text{if } robID \text{ takes care of it.} \end{cases} \tag{6}$$

We need to point out that AR_j^t is the accomplishment rate of $task_j$ at time t. Clearly, $AR_j^t \in [0, 1]$. We also define a new parameter called damaging rate function, labeled with $C_j(AR_j^t)$. It represents how much damages $task_j$ will cause to the system in per unit of time when the accomplishment rate equals to AR_j^t.

We use auction as a basic method to allocate tasks, a formalized auction can be represented as below:

$$A = < B, T > \tag{7}$$

B is the set of bidders, in other words, robots waiting for tasks. T is the set of tasks waiting to be assigned in this auction.

For a certain task t_j, bidder b_i will calculate two parameters: t_{ij}^1 and t_{ij}^2. t_{ij}^1 represents the time it takes for b_i to go to task t_j. t_{ij}^2 represents the time it takes for b_i from starting working on t_j to accomplishing it. So far, if t_j is assigned to b_i at time t_0, then under such allocation result, the damage that $task_j$ will cause to the system can be formalized as below:

$$cost_i^j = \int_{t_0}^{t_0+t_1^{ij}} C_j(0)dt + \int_{t_0+t_{ij}^1}^{t_0+t_{ij}^1+t_{ij}^2} C_j(AR_j^t)dt \tag{8}$$

4   Static Auction Algorithm Description

In this part, we consider one to one auction model of static task allocation with N robots matching N rescue tasks. The assignment problem to be discussed here is the mathematical problem of minimizing the damage of the whole system by matching N robots with N tasks. Constraint conditions of the algorithm can be expressed in a mathematical formula as below:

$$\min \quad \sum_{i \in B} \sum_{j \in T} x_{ij} C_{ij} \tag{9}$$

$$\text{s.t.} \quad \sum_{j|(i,j)\in T} x_{ij} = 1, \forall i = 1, 2, \ldots, n \tag{10}$$

$$\sum_{i|(i,j)\in T} x_{ij} = 1, \forall j = 1, 2, \ldots, n \tag{11}$$

$$x_{ij} = 0, 1, \forall (i, j) \in T \tag{12}$$

In this format group above, we use set T to represent the tuples of all possible distributions. It can be formalized as below:

$$T = \{(i, j)|j \in T(i), \forall i = 1, 2, \ldots, n\} \tag{13}$$

We use $T(i)$ to represent the set of all tasks that can be assigned to the robot i. In order to reduce the algorithm complexity, we use $\bar{t}$ to set a time limit, if the total time of robot i solving task j is more than t, task j cannot be assigned to robot i, which can be described as below:

$$T(i) = \{j|t_{ij} \leq \bar{t}\} \tag{14}$$

We use the set A to represent the two-tuples (i, j) consisting of robots and tasks. Each robot can have one two-tuples $(i, j) \in A$ at most. Each robot can have one two-tuples $(i, j) \in A$ at most either. For the set A, if there is a two-tuples$(i, j) \in A$, it means that task j is assigned to robot i.

In the algorithm, we set a positive value called ϵ and a price set $p = \{p_1, \ldots, p_n\}$. For robot i, if the difference between its absolute value of the relative gains obtained from task j and the optimal relative gains obtained from all the allocation schemes is not greater than ϵ, we call robot i and task j satisfy complementary slackness condition. This two-tuples (i, j) is the optimal result, which can be described as below:

$$\left|p_j - C_{ij} - \max_{k \in T(i)} \{p_k - C_{ik}\}\right| \leq \epsilon \tag{15}$$

The specific process of auction algorithm is described as follows.

Step 1 Select a $\epsilon > 0$, set $p_k = 0, \forall k = 1, 2, \ldots, n$, the set of robots that are not assigned tasks, is denoted as N, and the set of robots which are tender to the task j in the bidding phase is denoted as $B(j)$;

Step 2 This step is an iterative process:

– *Decision phase*: For each robot i in the set N, get the maximum relative gains u_i and the assigned task j_i when the maximum relative gain is obtained:

$$u_i = \max_{k \in T(i)} \{p_k - C_{ik}\} \tag{16}$$

And its second relative gains v_i:

$$v_i = \max_{k \in T(i), k \neq j} \{p_k - C_{ik}\} \tag{17}$$

If $T(i)$ has only one task j, we define v_i as $-\infty$.
– *Bidding phase*: All robots bid for j_i which is the most gainful task, the bidding price of the robot is determined as:

$$a_{ij_i} = p_{j_i} - u_i + v_i - \epsilon = C_{ij_i} + v_i - \epsilon \tag{18}$$

– *Allocation phase*: For each task j, if the $B(j)$ is not empty, we update the price of the task j to the highest bid price:

$$p_j = \max_{i \in B(j)} a_{ij} \tag{19}$$

The task is assigned to the highest bidding robot i_j, and at the same time, the two-tuples related to the robot i_j and the task j are removed from the A which is an infeasible allocation, and the new two-tuples (i_j, j) are added to the set A.

In this iterative process of the algorithm, we use p_j and p'_j, respectively, to represent the price corresponding to the task before and after the iteration. In the iterative process, if robot i bids to the task j_i and is successfully assigned to the task j_i, its price will be updated, which can be described as follows:

$$p'_{j_i} = C_{ij_i} + v_i - \epsilon \tag{20}$$

For each task j, there are $p_j \geq p'_{j_i}$, so we can get the format below:

$$|p'_{j_i} - C_{ij_i} - \max_{k \in T(i), k \neq j} \{p_k - C_{ik}\}| \leq \epsilon \tag{21}$$

It can be seen that after every iteration process, every two-tuples (i, j) always satisfies the complementary slackness condition.

5 Experimental Results and Future Work

In this paper, we propose a new auction model and use the auction algorithm to study multi-robot task allocation problem. Through experiments, we find that the task allocation can be effectively carried out by our algorithm.

We choose the classic Hungarian algorithm to do contrastive experiments. For all cost matrices, the same optimal assignment results can be obtained by iteration. The time spent on auction algorithm and Hungarian algorithm is as follows (see Fig. 1).

On the different scale of task allocation, the speed advantage of auction algorithm is obvious. The number of tasks has a greater impact on the running time of the Hungarian algorithm than the algorithm we proposed.

Due to the limited ability of the author, there are many shortcomings in this paper. On the one hand, new tasks may occur during the execution of a task, robots may need to stop their current work and carry out new tasks. On the other hand, a mechanism must be added to the auction algorithm to find out the infeasibility of the problem when the problem is insoluble. All of these are discussed in the future work.

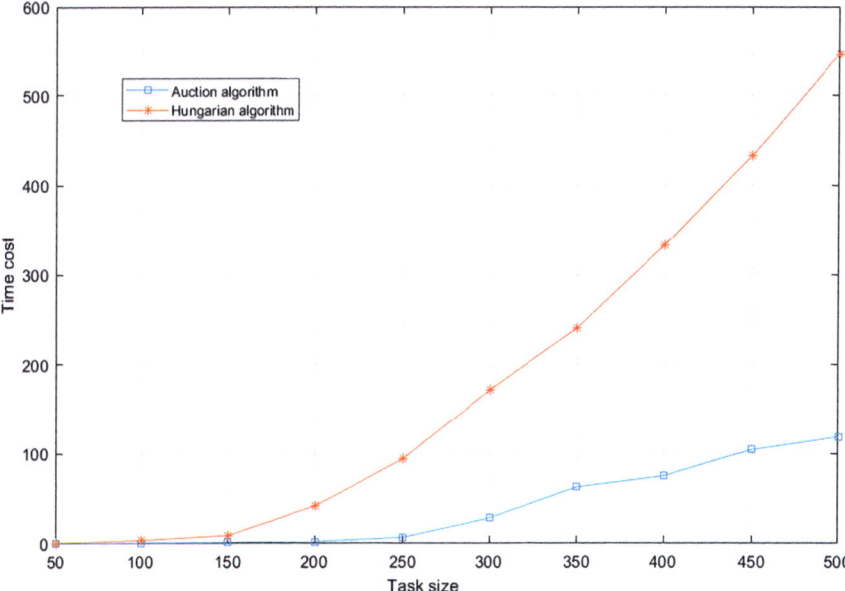

Fig. 1 Time cost comparison between auction algorithm and Hungarian algorithm

Acknowledgements This work was supported by the National Nature Science Foundation of China under Grant 61170201, Grant 61070133, and Grant 61472344, in part by the Innovation Foundation for graduate students of Jiangsu Province under Grant CXLX12 0916, in part by the Natural Science Foundation of the Jiangsu Higher Education Institutions under Grant 14KJB520041, in part by the Advanced Joint Research Project of Technology Department of Jiangsu Province under Grant BY201506106 and Grant BY2015061-08, and in part by the Yangzhou Science and Technology under Grant YZ2017288 and Yangzhou University Jiangdu High-end Equipment Engineering Technology Research Institute Open Project under Grant YDJD201707.

References

1. Booth, K. E., Nejat, G., & Beck, J. C. (2016). A constraint programming approach to multi-robot task allocation and scheduling in retirement homes. In *International Conference on Principles and Practice of Constraint Programming* (pp. 539–555). Berlin: Springer.
2. Das, G. P., McGinnity, T. M., Coleman, S. A., & Behera, L. (2015). A distributed task allocation algorithm for a multi-robot system in healthcare facilities. *Journal of Intelligent & Robotic Systems, 80*(1), 33–58.
3. Elango, M., Nachiappan, S., & Tiwari, M. K. (2011). Balancing task allocation in multi-robot systems using k-means clustering and auction based mechanisms. *Expert Systems with Applications, 38*(6), 6486–6491.
4. Garg, R., & Kapoor, S. (2006). Auction algorithms for market equilibrium. *Mathematics of Operations Research, 31*(4), 714–729.

5. Hooshangi, N., & Alesheikh, A. A. (2017). Agent-based task allocation under uncertainties in disaster environments: An approach to interval uncertainty. *International Journal of Disaster Risk Reduction, 24*, 160–171.

6. Jiang, L., & Zhang, R. (2011). An autonomous task allocation for multi-robot system. *Journal of Computational Information Systems, 7*(11), 3747–3753.

7. Liu, Y., Yang, J., Zheng, Y., Wu, Z., & Yao, M. (2013). Multi-robot coordination in complex environment with task and communication constraints. *International Journal of Advanced Robotic Systems, 10*(5), 229.

8. Lu, L., Yu, J., Zhu, Y., & Li, M. (2018). A double auction mechanism to bridge users' task requirements and providers' resources in two-sided cloud markets. *IEEE Transactions on Parallel and Distributed Systems, 29*(4), 720–733.

9. Ponda, S. S., Johnson, L. B., & How, J. P. (2012). Distributed chance-constrained task allocation for autonomous multi-agent teams. In: *American Control Conference (ACC), 2012* (pp. 4528–4533). Piscataway: IEEE.

10. Tang, J., Zhu, K., Guo, H., Gong, C., Liao, C., & Zhang, S. (2018). Using auction-based task allocation scheme for simulation optimization of search and rescue in disaster relief. *Simulation Modelling Practice and Theory, 82*, 132–146

11. Wang, L., Liu, M., & Meng, M. Q. H. (2017). A hierarchical auction-based mechanism for real-time resource allocation in cloud robotic systems. *IEEE Transactions on Cybernetics, 47*(2), 473–484.

Non-uniformity Detection Method Based on Space-Time Autoregressive

Ying Lu

1 Introduction

In recent years, AR model has been applied to multi-channel radar detection [1, 2]. PAMF, which is constructed by matrix whitening filter and residual filter, was proposed by Michels et al. by combining STAP with multi-channel parametric model in a creative way [3, 4]. In this method, vector linear predictive filter and residual filter whiten the input data, respectively, in time and space domain. Swindlehust et al. [2, 5] proposed STAR by improving the vector linear predictive filter furthermore. STAR can whiten input data in time and space domains at the same time only using vector linear predictive filter and decreasing dimension of space domain of vector linear predictive filter. Some methods have been proposed to apply to the field of imaging target recognition and signal processing [6–11].

STAR [12] estimates the parameters of transmission model based on vector autoregression model, and extrapolates the covariance matrix of clutter of the received data to overcome shortcoming of performance penalties due to the lack of training samples. STAR is a brand new method which is totally different from traditional STAP which reduces dimension and rank with a faster convergence speed and less requirement of samples.

In Ref. [13], they proposed an anti-interference target $\Sigma\Delta$ - STAR algorithm in the existence of inhomogeneous clutter power and interference target (outlier) only acquiring quite a few training samples. In addition, there is isolated interference which is also inhomogeneous in the realistic radar environment, and the lost information due to the isolated interference is unable to be obtained from the other training samples. The method proposed in Ref. [13] has a bad performance on target

Y. Lu (✉)
China Academy of Launch Vehicle Technology, Beijing, China

© Springer Nature Switzerland AG 2020
H. Lu, L. Yujie (eds.), *2nd EAI International Conference on Robotic Sensor Networks*, EAI/Springer Innovations in Communication and Computing,
https://doi.org/10.1007/978-3-030-17763-8_15

detection because of the isolated interference. Aimed at this case, we propose a new Cascade STAR method suppressing all three inhomogeneous interferences. Our method constitutes two steps: (1) The outlier-resistant method analogized from $\Sigma\Delta$ - STAR proposed in Ref. [13] is used to resist the interference target, meanwhile, cancel the negative influence of inhomogeneous clutter power with a few training samples. Then, we use the homogeneous samples to suppress the homogeneous clutter. (2) We process the data to be detected with STAR method to suppress the isolated interference.

In order to suppress all three kinds of inhomogeneous phenomena effectively of clutter power, interference target and isolated interference, this paper models and optimizes Cascade STAR method. The outline of this paper is as follows: In Sect. 2, we briefly review the basic principle of STAR algorithm, and then, Cascade STAR algorithm is described and detailed in Sect. 3. Simulation experiment results demonstrate that the proposed method has good performance, which is presented in Sect. 4. At last, the conclusions are given in Sect. 5.

2 The Basic Principle of Star Algorithm

Assume that the kth multi-dimensional echo data vector $X_l(k)$ received by the lth range unit follows AR model [14]:

$$\varepsilon_l(k) = \sum_{i=0}^{p-1} A_i X_l(k-i) \qquad k = 1, \cdots, K - p + 1 \tag{1}$$

where $\varepsilon_l(k)$ is the residual error vector, $A_i \in \mathbb{R}^{N' \times N}$ is the space-time autoregressive coefficient matrix, N is the spatial degree of freedom (DOF), $N' \leq N$ is the DOF after reducing dimensions, p is the rank of model and K is the pulse number of time domain. We define

$$A^H = [A_0, A_1, \cdots, A_{p-1}] \tag{2}$$

where superscript H denotes the conjugate transpose. Data of the lth range unit can be written as

$$e_l = \begin{bmatrix} X_l(1) & \cdots & X_l(K-p+1) \\ \vdots & \vdots & \vdots \\ X_l(p) & \cdots & X_l(K) \end{bmatrix} \tag{3}$$

then, all training samples can be denoted by

$$E = [e_1, e_2, \cdots, e_L] \tag{4}$$

where L is the number of training samples with the constraint $A^H A = I$ to obtain the nontrivial space-time autoregressive coefficient matrix. Then A_k is the left singular vector associated with the kth smallest singular value of sample matrix E. Thus, the number of samples is increased by $K - p + 1$ fold due to the smoothness of data. With Eqs. (2) and (3), and the definition (4), Eq. (1) can be written as

$$\varepsilon = A^H E = [\varepsilon_1, \varepsilon_2, \cdots, \varepsilon_L] \tag{5}$$

where ε_l is

$$\varepsilon_l = [\varepsilon_l(1), \varepsilon_l(2), \cdots, \varepsilon_l (K - p + 1)] \tag{6}$$

We set

$$A^H E = 0 \tag{7}$$

The STAR whitening filter can be represented by

$$\Gamma = \begin{bmatrix} A_0 \cdots A_{p-1} & 0 & 0 \\ 0 & \ddots & & \ddots & 0 \\ 0 & 0 & A_0 & \cdots A_{p-1} \end{bmatrix} \tag{8}$$

According to Ref. [14], diagonally loaded STAR which has a better capability of detection is more robust than prediction error STAR in the choice of the rank of space domain N'; thus, the weight vector of STAR is calculated as

$$W = \Gamma^H \left(\Gamma \Gamma^H + \kappa I \right)^{-1} \Gamma S \tag{9}$$

where S is the space-time steering vector of target signal, and κ is the diagonal loading factor. The number of training samples of STAR method is $L \geq [Np/K - p + 1]$ [2]. N' is calculated as

$$N' = \left\lceil \frac{NK - \rho}{K - p + 1} \right\rceil \tag{10}$$

by the rule in Ref. [2], and rank p is usually determined by

$$\text{AIC} \left(p' \right) = N_D \ln \left| \hat{\Sigma} \right| + 2 N_L \left(p' \right) \tag{11}$$

$$\text{MDL} \left(p' \right) = N_D \ln \left| \hat{\Sigma} \right| + N_L \left(p' \right) \ln \left(N_D \right) \tag{12}$$

which is inspired by that the ROF of system after dimensional reduction is not smaller than the summary of dimensions of the ROF of clutter and the spatial dimension of noise after rank reduction, then, the rank p can be calculated as

$$Np \geq \rho + N' \tag{13}$$

where $N_D = 2L(K - 1)$, $\hat{\Sigma} = \text{AEE}^H A^H / [L(K - p'], N_L(p') = 4p' - 3$ and ρ is the DOF of the clutter. The amount of information obtained by MDL method has statistic consistency; however, the rank obtained by AIC method will be biased when pulse trains are short. Wu et al. [15] improve the rank rule as

$$p = \left\lceil \frac{\rho + N'}{N} \right\rceil \tag{14}$$

Therefore, STAR is a dimensional and rank reduction method decreasing computation and the requirement of amount of samples. More importantly, it increases the number of samples by smoothing the data to have a good performance in the inhomogeneous environment during the process of dimensional reduction.

3 Cascade Star Algorithm

3.1 Anti-Interference Target STAR Algorithm

In [13], Shen et al. proposed an approach for determining the interference target's Doppler frequency based on the local maximum value of the weight vector. This idea is aimed at cancelling some particular signals with Doppler frequency f_1, then utilizing STAR for the cancelled data to obtain a weight vector W_1. When the Doppler frequency of cancellation weight coefficient is equivalent to the Doppler frequency of interference target, we set

$$\|W_1\| > 0 \tag{15}$$

Otherwise, we set

$$\|W_1\| = 0 \tag{16}$$

Therefore, we can cancel the signal by selecting different Doppler frequencies around the Doppler frequency to be detected, and calculate the associated weight vector with STAR to determine the Doppler information of outliers. If there are outliers existing in the search scope, the Doppler frequency associated with the maximum value of norm of weight vector is the Doppler frequency of interference target. We first introduce the anti-interference target STAR method on the basis of

Ref. [13]. Assume that the antennas are composed of linear uniform array with N elements, and the number of pulses in a CPI is K, then the snapshot data received by the lth range unit is

$$
X_l =
\begin{bmatrix}
X_l(1,1) & X_l(1,2) & \cdots & X_l(1,K) \\
X_l(2,1) & X_l(2,2) & \cdots & X_l(2,K) \\
\vdots & \vdots & \vdots & \vdots \\
X_l(n,1) & X_l(n,2) & \cdots & X_l(n,K) \\
\vdots & \vdots & \vdots & \vdots \\
X_l(N,1) & X_l(N,2) & \cdots & X_l(N,K)
\end{bmatrix}
\tag{17}
$$

We choose a Doppler frequency f_1 around the Doppler frequency to be detected, and weight-cancel it with $z_1 = e^{-j2\pi \frac{f_1}{f_r}}$ (f_r is the pulse repetition frequency), then the data after signal cancellation is

$$
Y_{l0} =
\begin{bmatrix}
X_l(1,1) & X_l(1,2) & \cdots & X_l(1,K-1) \\
X_l(2,1) & X_l(2,2) & \cdots & X_l(2,K-1) \\
\vdots & \vdots & \vdots & \vdots \\
X_l(n,1) & X_l(n,2) & \cdots & X_l(n,K-1) \\
\vdots & \vdots & \vdots & \vdots \\
X_l(N,1) & X_l(N,2) & \cdots & X_l(N,K-1)
\end{bmatrix}
$$

$$
-
\begin{bmatrix}
X_l(1,2) & X_l(1,3) & \cdots & X_l(1,K) \\
X_l(2,2) & X_l(2,3) & \cdots & X_l(2,K) \\
\vdots & \vdots & \vdots & \vdots \\
X_l(n,2) & X_l(n,3) & \cdots & X_l(n,K) \\
\vdots & \vdots & \vdots & \vdots \\
X_l(N,2) & X_l(N,3) & \cdots & X_l(N,K)
\end{bmatrix} \cdot z_1
\tag{18}
$$

then Y_{l0} can be written as

$$
Y_l =
\begin{bmatrix}
Y_l(1) & \cdots & Y_l(K-p) \\
\vdots & & \vdots \\
Y_l(p) & \cdots & Y_l(K-1)
\end{bmatrix}
\tag{19}
$$

where $Y_l(k)$ is the kth column vector of Y_{l0}. All samples can be written in a matrix form as

$$
E = [Y_1 \ Y_2 \ \cdots \ Y_L]
\tag{20}
$$

We can get more samples if the original data is conjugate transposed in the opposite direction, and the weight vector W_1 can be obtained by training the data cancelled with z_1 according to Eqs. (2), (8) and (9), Then norm of W_1 is calculated as

$$W_1 = \Gamma_1^H \left(\Gamma_1 \Gamma_1^H + \kappa I \right)^{-1} \Gamma_1 S_1 \tag{21}$$

where S_1 is a space-time steering vector associated with f_1. Based on the above introduction, we can calculate the weight vectors associated with the cancelled signals which are chosen from the vicinity of the Doppler frequency to be detected. Whether the interference targets exist around the Doppler frequency to be detected depends on the norm of the associated weight vector. We can filter the interference targets with Eq. (18) if they exist. Analysis of the algorithm shows that the time domain ROF will reduce by one when an interference target is filtered. However, the decrease of time domain ROF only leads to the reduction of samples, which can be dealt with increasing the number of samples.

3.2 Cascade STAR Algorithm

With the training samples without interference targets which are filtered by the anti-interference target STAR introduced in the previous section, the weight value we obtained can suppress the information of homogeneous clutter in the unit to be detected. Though, there is isolated interference in the realistic environment, the Cascade STAR method can efficiently deal with that. However, the anti-interference target STAR method is unable to suppress the isolated interference. The detailed procedure of our Cascade STAR is as follows: Since STAR method only needs the data from quite a few range units, it avoids from the nonhomogeneity of clutter power from the perspective of algorithm; Resist interference targets: Interference targets are filtered by the anti-interference target STAR method. Suppress the homogeneous clutter in the unit to be detected: W_1 obtained from the data without interference targets which are denoted by $Y_l (l = 1, 2, \cdots, L)$ can be used to suppress the homogeneous clutter in the unit to be detected. The weight value obtained from the samples Y_l without the information of interference target has no effect on the isolated interference as the result of the different distribution characteristics between the isolated interference and the homogenous clutter in beam-Doppler space; Suppress isolated interference: To avoid signal cancellation, we first process the target signal as

$$\|W_1\| = \left[S_1^H \Gamma_1^H \left(\Gamma_1 \Gamma_1^H + \kappa I \right)^{-1} \Gamma_1 \Gamma_1^H \left(\Gamma_1 \Gamma_1^H + \kappa I \right)^{-1} \Gamma_1 S_1 \right]^{1/2} \tag{22}$$

$$Y_0 = \begin{bmatrix} X_0(1,1) & X_0(1,2) & \cdots & X_0(1,K-1) \\ X_0(2,1) & X_0(2,2) & \cdots & X_0(2,K-1) \\ \vdots & \vdots & \vdots & \vdots \\ X_0(n,1) & X_0(n,2) & \cdots & X_0(n,K-1) \\ \vdots & \vdots & \vdots & \vdots \\ X_0(N-1,1) & X_0(N-1,2) & \cdots & X_0(N-1,K-1) \end{bmatrix}$$

$$-\begin{bmatrix} X_0(1,2) & X_0(1,3) & \cdots & X_0(1,K) \\ X_0(2,2) & X_0(2,3) & \cdots & X_0(2,K) \\ \vdots & \vdots & \vdots & \vdots \\ X_0(n,2) & X_0(n,3) & \cdots & X_0(n,K) \\ \vdots & \vdots & \vdots & \vdots \\ X_0(N-1,2) & X_0(N-1,3) & \cdots & X_0(N-1,K) \end{bmatrix} \cdot z_t \cdot z_s \qquad (23)$$

where $z_t = e^{-j\pi f_0}$, $z_s = e^{-\frac{2\pi d}{\lambda}\cos\theta_0\cos\varphi_0}$, f_0 is the normalized Doppler frequency of the target signal and θ_0, φ_0 are the azimuth and pitch angle of the target angles, respectively. A new Y_0^* sample can be obtained by conjugate transposing Y_0. We write Y_0 and Y_0^* in the form of Eq. (20) and still denote them as Y_0 and Y_0^* for simplicity. Then, the matrix of the training samples which suppress the isolated interference is

$$E_{\mathrm{DI}} = \begin{bmatrix} Y_0 & Y_0^* \end{bmatrix} \qquad (24)$$

We denote $H_i(i = 1, 2, \cdots, p-1) \in \mathbb{R}^{N' \times N}$ as the matrix of space-time regression coefficient and

$$H^H = \begin{bmatrix} H_0, H_1, \cdots, H_{p-1} \end{bmatrix} \qquad (25)$$

We set

$$H^H E_{\mathrm{DI}} = 0 \qquad (26)$$

to avoid trivial solution with constraint $H^H H = I$, where H is the left singular vector associated with the corresponding value of N' minimum singular values of E_{DI}. We set

$$\Gamma = \begin{bmatrix} H_0 & \cdots & H_{p-1} & 0 & 0 \\ 0 & \ddots & & \ddots & 0 \\ 0 & 0 & H_0 & \cdots & H_{p-1} \end{bmatrix} \qquad (27)$$

then the associated weight vector is calculated as

$$W = \Gamma^{H}\left(\Gamma\Gamma^{H} + \kappa I\right)^{-1}\Gamma^{H}W_{1} \tag{28}$$

The flowchart of Cascade STAR algorithm is shown in Fig. 1.

4   Simulation Results and Analysis

The training sample E_{DI} contains the information of isolated interference and homogeneous clutter. However, we can consider that there are no information of homogeneous clutter in E_{DI} but information of isolated interference in consequence of W_1 in Eq. (9) replacing the space-time steering vector in Eq. (28) when we consider that W_1 is equivalent to the space-time steering vector. Because of the suppression of most of the homogeneous clutter and noise in the unit to be detected and a few requirements of samples of STAR, our method is still efficient in suppressing the isolated interference when only two training samples are utilized in the experiment. We take some experiments to verify the performance of our Cascade STAR method. In our experiment, the side-looking multi-channel radar system has eight rows and eight columns, the number of pulses in a CPI is 10, the radar wavelength is 0.2 m and the pulse repetition frequency is 3000 KHz. The input target signal and interference parameters are shown in Table 1 and Fig. 2.

Figure 3 shows the norm of weight vector of the normalized Doppler frequency ranged from 0.3 to 0.5 of the target's cancellation weight coefficient. Our method can estimate the Doppler frequency of interference target correctly, aka, the anti-interference target STAR can filter the interference signals. Figure 4a, b shows the training weight value obtained from anti-interference target method and Cascade STAR, respectively.

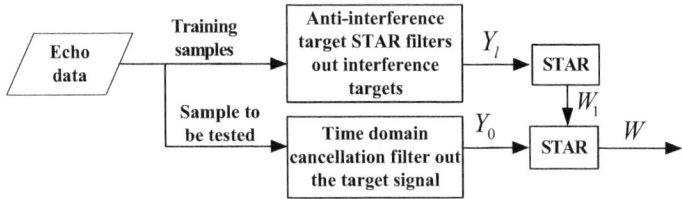

Fig. 1 The flowchart of cascaded SATR algorithm

Table 1 Input target signal and interference parameters

	Distance unit number	Power (dB)	Azimuth	Normalized Doppler frequency ($2f_d/f_r$)
Target signal	30	20	90°	0.4
Isolated interference	30	20	90°	−0.2
Interference target	28, 32	40, 40	90°	0.45, 0.36

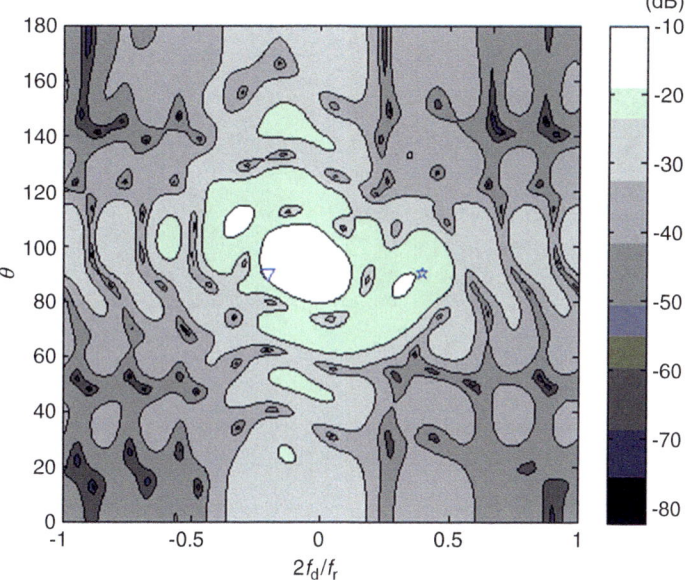

Fig. 2 Parameters of input signal

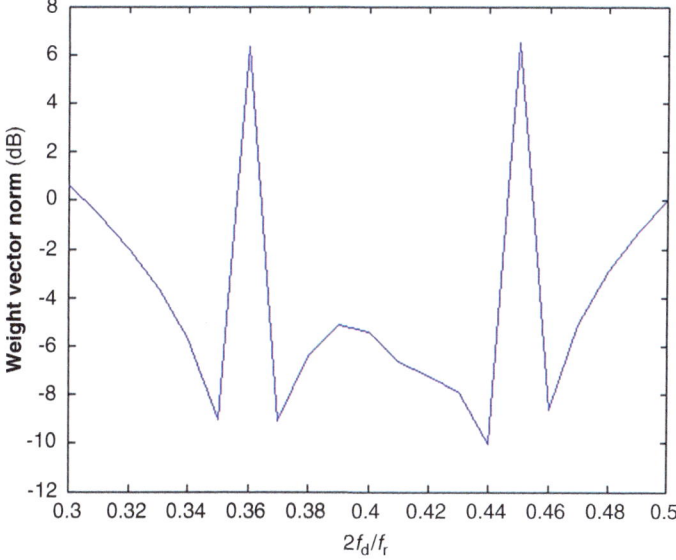

Fig. 3 Determination of Doppler frequency of outlier

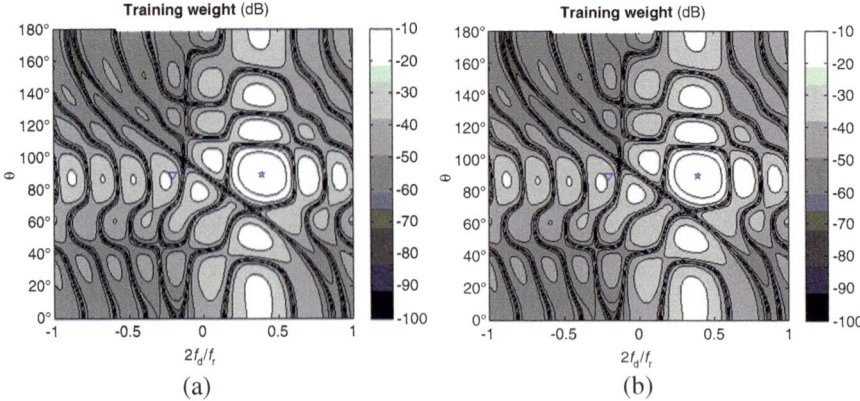

Fig. 4 Training weight value of anti-interference target STAR and Cascaded STAR. (**a**) Training weight of Anti-interference target STAR. (**b**) Training weight of Cascaded STAR

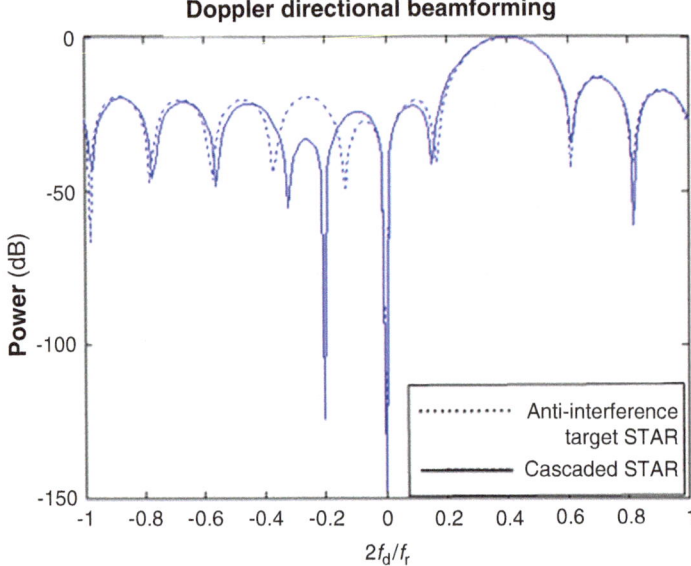

Fig. 5 Comparison between beam-forming diagrams in the Doppler direction of our method and anti-interference target STAR

Figure 4a, b shows that both methods have a good performance on suppressing clutter as there are deep depressions in the clutter areas, but our method has a deeper depression than anti-interference target method. Figure 5 shows this advantage more clearly in the form of beam-forming diagram in the Doppler direction. We can see that our method has a sufficient depression in the isolated interference direction; however, anti-interference target STAR does not. Therefore, our method has a better performance in inhomogeneous environment.

Theoretic analysis and experiment results show that our method which takes full advantages of anti-interference target STAR efficiently avoids from inhomogeneous clutter power only acquiring a few training samples. It combines anti-interference target STAR with the original STAR method with the training samples which filter out target signals and not only resist interference targets but also suppress the isolated interference sufficiently. Thus, our method has a good capability of suppressing inhomogeneous clutter power, interference targets and isolated interference in inhomogeneous environment.

5 Conclusions

In this paper, we have introduced STAT method, then propose a Cascade STAR method to improve the effect of STAR which has a bad performance in inhomogeneous environment on suppressing the isolated interference. Through theoretic analysis and experiment results, the method is proved to have a better performance in suppressing isolated interference and is more suitable in inhomogeneous environment.

References

1. Roman, J. R., Ranga, M., Davis, D. W., et al. (2000). Parametric adaptive matched filter for airborne radar applications. *IEEE Transactions on Aerospace and Electronic Systems, 36*(2), 677–692.
2. Parker, P., & Swindlehurst, A. (2003). Space-time autoregressive filtering for matched subspace. *IEEE Transactions on Aerospace and Electronic Systems, 4*(2), 510–520.
3. Michels, J. H. (1995). Multichannel signal detection involving temporal cross-channel correlation. *IEEE Transactions on Aerospace and Electronic Systems, 10*(3), 866–880.
4. Roman, J. R., Rangaswamy, M., Davis, D. W., et al. (2000). Parametric adaptive matched filter for airborne radar. *IEEE Transactions on Aerospace and Electronic Systems, 36*(2), 677–692.
5. Swindlehurst, A. L., & Parker, P. (2000). Parametric clutter rejection for space-time adaptive processing. In *Proceedings of the ASAP Workshop*. Lexington: MIT Lincoln Lab.
6. Lu, H., Li, Y., Chen, M., Kim, H., & Serikawa, S. (2018). Brain intelligence: Go beyond artificial intelligence. *Mobile Networks and Applications, 23*(2), 368–375.
7. Li, Y., Lu, H., Li, J., Li, X., Li, Y., & Seiichi, S. (2016). Underwater image de-scattering and classification by deep neural network. *Computers and Electrical Engineering, 54*, 68–77.
8. Li, Y., Lu, H., Li, K.-C., Kim, H., & Serikawa, S. (2017). Non-uniform de-scattering and de-blurring of underwater images. *Mobile Networks and Applications, 23*(2), 352–362.
9. Deng, L., Zhu, H., Zhou, Q., & Li, Y. (2018). Adaptive top-hat filter based on quantum genetic algorithm for infrared small target detection. *Multimedia Tools and Applications, 77*(9), 10539–10551.
10. Deng, L., & Zhu, H. (2016). Infrared moving point target detection based on spatial-temporal local contrast filter. *Infrared Physics & Technology, 76*, 168–173.
11. Deng, L., & Zhu, H. (2015). Moving point target detection based on clutter suppression using spatial temporal local increment coding. *Electronics Letters, 51*(8), 625–626.

12. Wu, D., Zhu, D., Shen, M., & Zhu, Z. (2012). Time-varying space-time autoregressive filtering algorithm for space-time adaptive processing. *IET Radar, Sonar and Navigation, 4*(6), 213–221.

13. Shen, M. (2008). *Research on moving target detection technology for heteroscedastic beam space-time processing*. Nanjing University of Aeronautics and Astronautics Doctoral Dissertation.

14. Russ, J. A., Casbeer, D. W., & Swindlehurst, A. L. (2004). STAP detection using space-time autoregressive filtering. In *Proceedings of the 2004 IEEE Radar Conference (IEEE Cat. No. 04CH37509)* (pp. 541–545). IEEE.

15. Wu, B. (2007). *Research on STAP technology of phased array airborne radar in heterogeneous clutter environment*. National University of Defense Technology Doctoral Dissertation.

Secondary Filter Keyframes Extraction Algorithm Based on Adaptive Top-K

Yan Fu, Chunlin Xu, and Mei Wang

1 Introduction

With the development of coal technology, the proportion of the coal industry in China is increasing, although the topic of the safety of coal production has also caused widespread concern in the community. National regulatory authorities of coal mines, in response to this problem, has opened a special National Coal Mine Safety Supervision Bureau, to deal with coal mine safety problems. Video surveillance plays an important role in the safety of coal mine production. It can send information about coal mine jobs to the production managers promptly, to guide coal mine production in an orderly fashion and to resolve dangerous situations in a timely manner. However, owing to the special environment of the coal mine, in the production of coal-based machinery, which makes the monotonous images obtained by the monitoring video and a little useful information, the further analysis and utilization of the image is limited [1]. Therefore, it is of great importance to extract the keyframes from the mine surveillance images.

In the next section, related works on keyframe extraction are introduced; in the third section, "Secondary Filter Keyframes Extraction Algorithm" based on eigenvalues; in the fourth section, "Experimentation and Analysis"; finally, the chapter ends with the "Conclusion."

Please note that the LNICST Editorial assumes that all authors have used the western naming convention, with given names preceding surnames. This determines the structure of the names in the running heads and the author index.

Y. Fu (✉) · C. Xu · M. Wang
School of Computer Science and Technology, School of Electrical and Control Engineering,
Xi'an University of Science and Technology, Xi'an, Shaanxi, People's Republic of China

© Springer Nature Switzerland AG 2020　　　　　　　　　　　　　　169
H. Lu, L. Yujie (eds.), *2nd EAI International Conference on Robotic Sensor Networks*, EAI/Springer Innovations in Communication and Computing,
https://doi.org/10.1007/978-3-030-17763-8_16

2   Related Work

Keyframe extraction can reduce the amount of redundant information content that exists between video frames. The information contained in a video can also be expressed succinctly to facilitate indexing and management of the video content, as shown in Fig. 1.

The low intensity of the illumination, the large amount of dust, and strong light interference in the coal mine lead to problems with blurring of the monitoring video and easily missing of target detection [2]. This has a great impact on information extraction from the coal mine video. If there is human interference, the accuracy of the keyframe extraction becomes worse.

For the keyframe extraction of images, there are many mature technologies, both at home and abroad, including the keyframe extraction methods based on shots division, clustering, motion analysis, and color features. Among them, the most widely used method is clustering-based processing, which divides each frame by setting a cluster center, and each cluster has a frame closest to a cluster center as a keyframe. Momin et al. [1] proposed an algorithm based on color features to calculate the value of the correlation through the color feature and then compare the value with the threshold to obtained keyframes. Lin and Lian [3] put forward the motion analysis method by calculating the motion energy of each frame, and then extracting the keyframe by comparing motion energy with the threshold. Both of these methods require a threshold to be set, and artificial threshold settings increase the contingency and instability of the method. Guan et al. [4] proposed a key point matching method based on a sliding window to get a similar frame, and finally extract the keyframe. The method is not good for tracking the key point of the environment when there is a loud noise in the mine. Sharma and Sathish [5] proposed a new method combining shots and clustering. After the initial shot segmentation, the keyframe selection is executed and the segmented shots are clustered. This method cannot meet the requirement of the structured information obtained in the case of a single environment in the mine.

In view of the above threshold-setting problem based on Momin et al. [1] involving correlation, this chapter is based on the research into the scale-invariant feature transform (SIFT) algorithm, the proposed adaptive threshold algorithm,

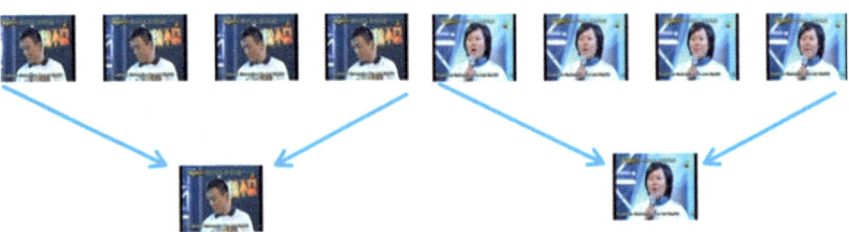

Fig. 1 Keyframe extraction

an adaptive method based on the frequency domain space calculation threshold, thereby reducing the influence of the human factor, and maintaining the time sequence of keyframes. The experimental results show that this algorithm improves the adaptability of the environment. Compared with the relevance-based keyframe extraction algorithm proposed by Momin et al. [1], the keyframe extracted by this algorithm has greater accuracy and better adaptability to the coal mine environment.

3   Secondary Filter Key Frame Extraction Algorithm

The method of keyframe extraction shows that redundant frames need to be deleted. Redundant frames are similar to keyframes; thus, each frame is calculated to express the value of the frame, which is mapped to the frequency domain to calculate the video frames as the keyframes within the threshold range.

Therefore, this chapter combines the methods of conference [1] based on the correlation of color space. The improved coal mine video keyframe extraction method proposed mainly includes: (1) video sequences of moving target detection; (2) calculation of the image features of the video frames; (3) adaptive threshold calculation; (4) a secondary filter to extract keyframes.

3.1   Video Sequence Moving Target Detection

Aimed at the problem of moving object detection, such as ray change, noise, and local motion, a moving object detection algorithm based on background subtraction [6] is proposed. The background subtraction method uses the parameter model of the background to approximate the image. It compares the current frame with the background image to realize the detection of the moving target. The algorithm focuses on building a robust background model to detect moving objects. It can adapt to the change in illumination, the movement of small targets in the background, and the influence of noise. The background subtraction method, Eq. (1), shows that:

$$\mathrm{BD}_t\,(x,\,y) = \begin{cases} 255, \left|I_t\,(x,\,y) - B_t\Big(x,\,y\Big)\right| \geq \tau_1 \\ 0, \left|I_t\,(x,\,y) - B_t\Big(x,\,y\Big)\right| < \tau_1 \end{cases} \tag{1}$$

where $I_t(x, y)$, $B_t(x, y)$ represent the current frame and the background frame image, $\mathrm{BD}_t(x, y)$ is the background difference image; t represents frames ($t = 1, 2, \ldots n$); τ_1 is the threshold.

3.2 Image Eigenvalue Calculation of Video Frames

The SIFT algorithm [7, 8] combines the scale invariant feature sub and gradient direction descriptor, so that it can maintain invariance with regard to rotation, scaling, and brightness changes, and a certain degree of stability to view change, affine transformation, and noise. Each key point extracted contains location, scale, and direction information.

The scale space of the two-dimensional image $I(x, y)$ is $L(x, y, \sigma)$, the convolution of a change scale Gauss equation $G(x, y, \sigma)$, and the original equation $I(x, y)$ are shown in Eq. (2).

$$L(x, y, \sigma) = G(x, y, \sigma) * I(x, y) \tag{2}$$

where x, y is the coordinates of P_{value}. σ is the factor of P_{value}, which is not involved in the calculation.

After the principal component analysis (PCA) dimensionality reduction, the matrix of the $1*n$ dimension of the C'_y is obtained, based on Eq. (3).

$$G(x, y) = \frac{1}{2\pi\sigma^2} e^{-\frac{(x-m/2)^2+(y-n/2)^2}{2\sigma^2}} \tag{3}$$

where m, n, σ is the factor of P_{value}, not involved in the calculation.

Equations (4) and (5) show the feature value of the video image, P_{value} is calculated by 2 norm. The image eigenvalue curve is shown in Fig. 2.

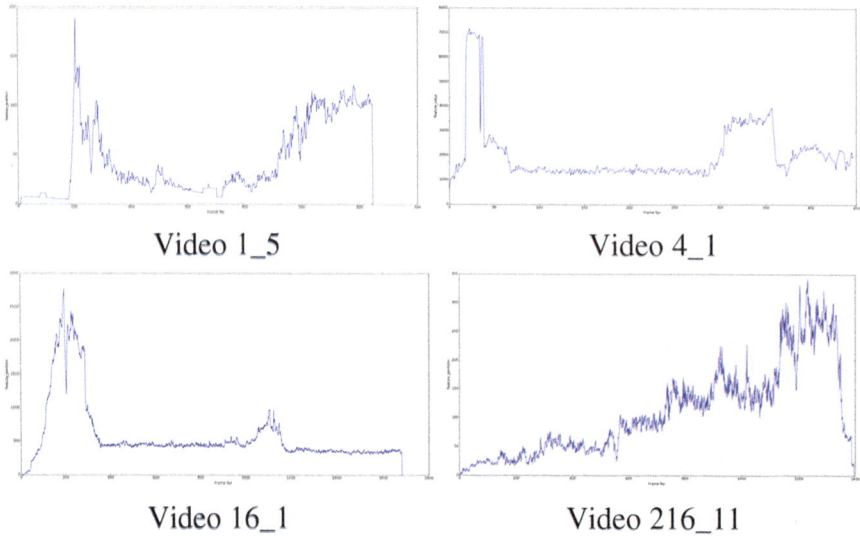

Video 1_5   Video 4_1

Video 16_1   Video 216_11

Fig. 2 Image eigenvalues curve, where the x-axis represents the amount of video frames, and the y-axis represents the size of the eigenvalues

$$C_y' = \begin{bmatrix} C_{1,1}' \\ C_{2,1}' \\ \vdots \\ C_{n,1}' \end{bmatrix} \tag{4}$$

$$P_{\text{value}} = \left\| C_y' \right\|_2 \tag{5}$$

where C_y' is the value feature vector of the video image dimension reduction by PCA.

3.3 Adaptive Threshold Calculation

Based on the eigenvalues, using a differential operator equation on P_{value} from the obtained $F(x_k)$, there is the first-order forward difference for the equations $f(x_k)$ shown in Eqs. (6)–(8).

$$f(x_k) = P_{\text{value}}(k), \ (k = 1, 2, 3, \ldots, n) \tag{6}$$

$$\Delta f(x_k) = f(x_{k+1}) - f(x_k) \tag{7}$$

$$F(x_k) = \Delta f(x_k) \tag{8}$$

where k is the P_{value}.

The equation $F(x_k)$ after the difference begins counting the distance from $F(x_k)$ to $f_{m\_n}(x, y)$, Eqs. (9) and (10). $f_{m\_n}(x, y)$ is the equation of the connection between the maximum and the minimum, and the $F(x_k)$ at this time is the threshold value.

$$f_{m\_n}(x, y) = r^* x + \min F(x) \tag{9}$$

$$r = \frac{\max F(x) - \min F(x)}{n} \tag{10}$$

where r is slope between the maximum and the minimum. $\max F(x)$, $\min F(x)$ are the maximum value of $F(x)$ and the minimum value of $F(x)$. $f_{m\_n}(x, y)$ is the equation connecting the maximum and the minimum of $F(x)$.

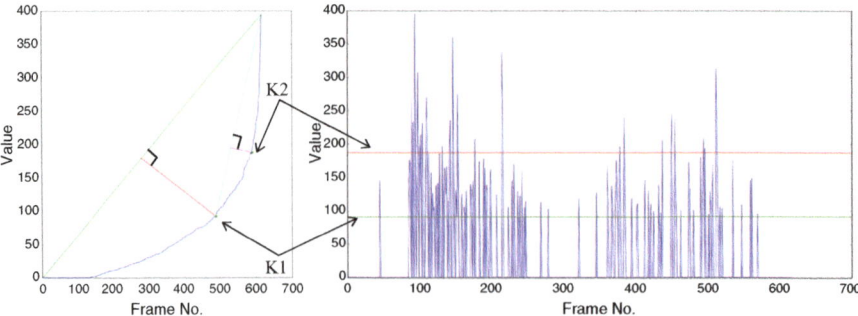

Fig. 3 Adaptive threshold calculation, where the x-axis represents the number of video frames, and the y-axis represents the size of the eigenvalues. On the left of the picture is P_{value} sorted by Top-K. On the right of the picture is the equation of difference $F(x)$. $k1$ is the threshold of the first filter; it can filter out more than 70% of the value. $k2$ is the threshold of the secondary filter; it represents the remaining keyframes after filtering

Equations (11) and (12) show the distance between $F(x_k)$ and $f_{m\_n}(x, y)$. The maximum value l_{max} is the distance from the point (x, y) on $F(x_k)$ to $f_{m\_n}(x, y)$. The $F(x_k)$ at this time is used as an adaptive threshold K. The adaptive threshold calculation is shown in Fig. 3.

$$L = \frac{|F(x_k) - r^* x_k - \min F(x)|}{\sqrt{1^2 + r^2}} \tag{11}$$

$$l_{\text{max}} = \max L \tag{12}$$

where, x_k, $x_k{'}$ is the value of P_{value}. $f(x_k)$ is the equation of the P_{value}. n is the amount of P_{value}, r is the slope between the maximum and the minimum. L is the distance value.

3.4 Secondary Filter Keyframe Extraction

A classification algorithm based on Top-K. In this chapter, using the quick sort Top-K classification is thought to sort the Eq. $F(x)$, and the first value is selected as the comparison object. If the elements of the array are less than p, then put the value on the left of p. If the elements of the array are greater than p, then put the value on the right of p. There is recursion to an orderly sequence of the entire array. In this chapter, we need to adjust the array to an ascending sequence from small to large; the sorting process is shown in Fig. 4.

Fig. 4 The sorting process

In the first filter, Eq. (13), $F(x)$ is divided into two classes with adaptive threshold $K = k1$. $k1$ is the threshold of the first filter, and it can filter out more than 70% of the value.

$$F(x) = \begin{cases} \Delta f(x_k), \ x > k1 \\ 0, \ x \leq k1 \end{cases} \quad n^*75\% \leq k1 \leq n^*95\% \tag{13}$$

where $k1$ is the first adaptive threshold. The range of $k1$ under normal conditions is $n^*75\%$ to $n^*95\%$. As some extreme data sets may not be in this range of $k1$, then the range of the $k1$ needs to be limited to continue the secondary filter. If $k1$ is less than 75%, there will be too many redundant frames, and if greater than 95%, there will be loss of effective keyframes.

Equation (14) shows the secondary filtration process for the remaining $n\text{-}x_k$ eigenvalues through the first filtering method, and then secondary filtration is calculated through the adaptive threshold $K = k2$. With the eigenvalues of video frames larger than $k2$, the first frame and the last frame are the keyframes. If there are continuous frames in the keyframe, only the first frame of the continuous frames is retained.

$$F(x) = \begin{cases} \Delta f(x_k), \ x > k2 \\ 0, \ x \leq k2 \end{cases} \tag{14}$$

where $k2$ is the second adaptive threshold. The secondary filter keyframe extraction method is shown below.

```
Input    :Video Frame :Fi    i=1,2,3,...,n
Output   :Keyframe: KFk    K=1,2,...,p(p<=n)
Terms    :D=Difference of two comparing frames
          T=Threshold(k1,k2. Calculation by the 2.3
chapter)
Initialisation : Use the frame as the keyframe that
corresponding to the threshold
              KFk=Fa    Fa=k1 or k2
  begin
    for i=a to n do
      if(i<=n) then
          D= Frame Difference(Fi,KFk)
      else
          stop
      end if
```

```
      if(D<=T)then
      k=k+1
      end if
      i=i+1
   end for
end
```

4   Experimentation and Analysis

To verify the effectiveness of the algorithm, this chapter compares several other mainstream algorithms and evaluates the effectiveness of each algorithm by using the Hautière evaluation index. Among them, the representative correct amount of the accuracy rate is divided by the extracted amount, and the representative correct amount of the recall rate is divided by the effective keyframes amount. Triangles represent redundant frames and the circle represents the wrong frame, as shown in Fig. 5. The integrity and accuracy of the algorithm in this chapter are better than for the other two algorithms. Tables 1 and 2 show the data of the three algorithms. The accuracy rate and recall rate of this algorithm are higher than those of other algorithms, and the effectiveness of the algorithm in this paper is proven. Experimental results are shown in Fig. 5.

5   Conclusion

The particularity of the coal mine environment leads to the incomplete extraction of keyframes by the methods of Lin and Lian [3]. In this chapter, an improved algorithm is proposed on the basis of the correlation based on Momin et al. [1]. First, the background subtraction method is used to extract the moving target. Using the improved correlation value algorithm, the feature points are extracted by SIFT [9], and the correlation value of the image is obtained by using the 2 norm of PCA to reduce the dimension. In the setting of the threshold, the adaptive threshold of the Top-K is calculated. Experimental results show that the algorithm can extract keyframes effectively and accurately, and achieves an adaptive purpose in the threshold setting, which reduces human interference.

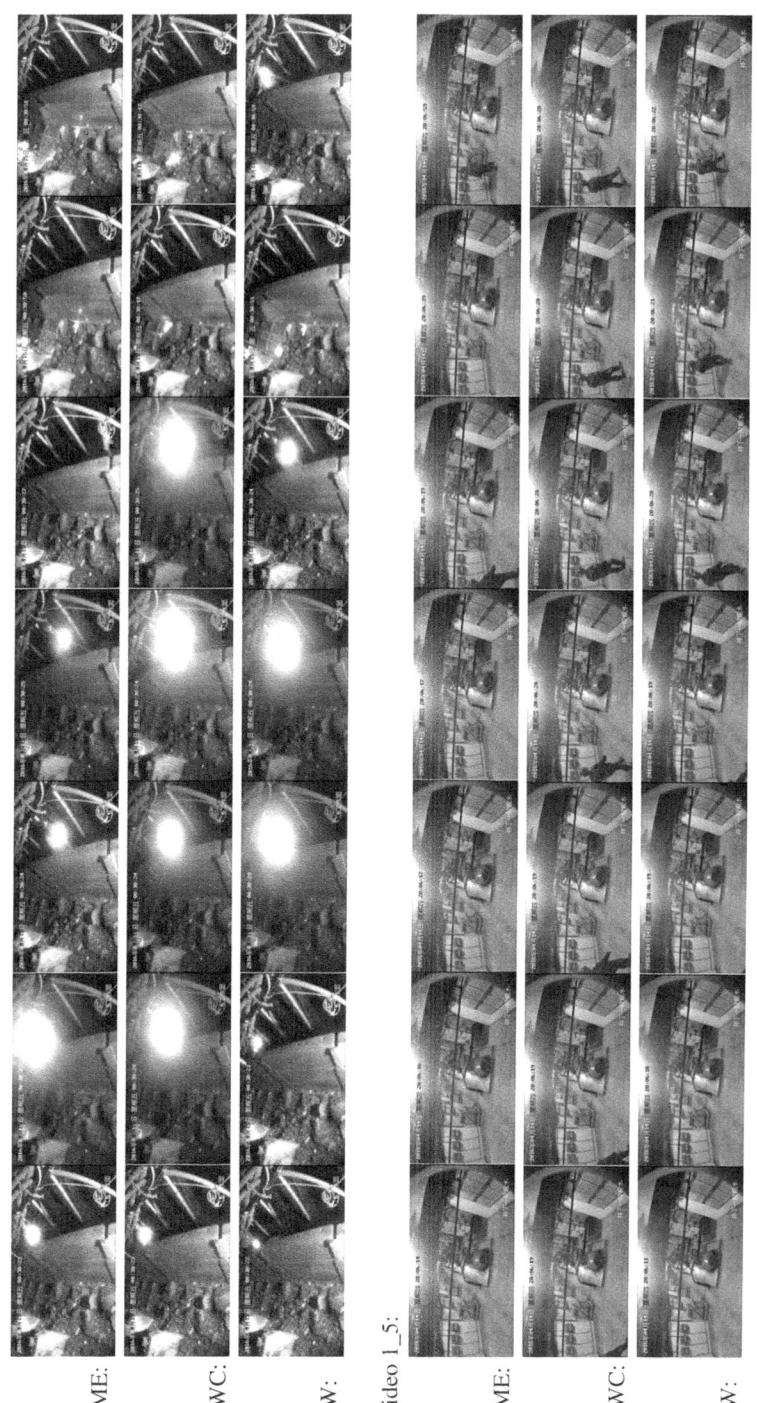

Fig. 5 Experimental results, where performance motion extraction (PME) is the algorithm of perceived motion energy [3], surveillance with content (SWC) is the algorithm based on correlation [1], FW is the algorithm of this chapter based on the feature value

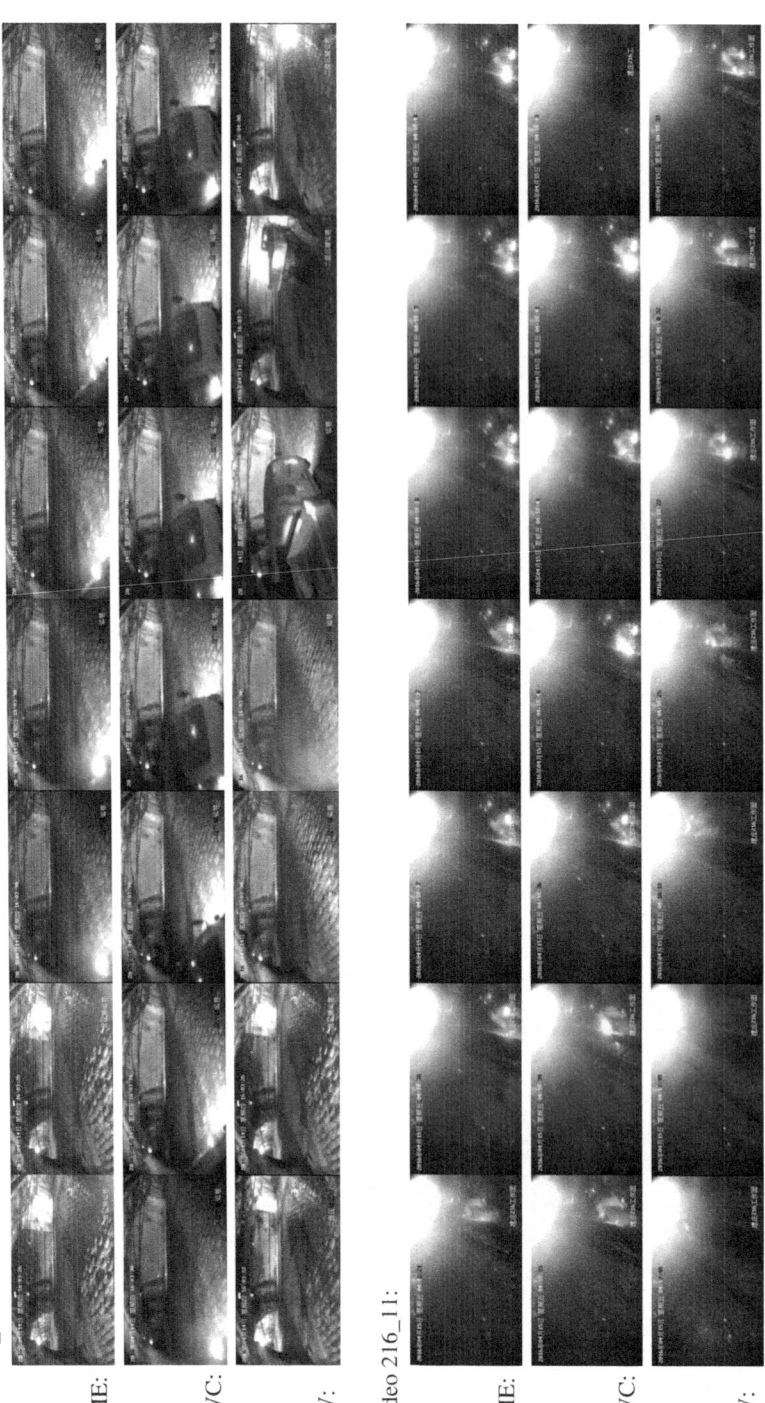

Video 16_1:

PME:

SWC:

FW:

Video 216_11:

PME:

SWC:

FW:

Fig. 5 (continued)

Table 1 Experimental data information

Video	Total frames	Number of shots	Number of effective keyframes
Video 4_1	448	3	8
Video 1_5	622	3	18
Video 16_1	1683	5	14
Video 216_11	1396	2	11

Number of effective keyframes means the number of effective keyframes that can express video

Table 2 Three keyframe algorithm extraction results

Video	Method used	Number extracted	Number correct	Number missed	Number wrong	Number redundant	Accuracy rate (%)	Recall rate (%)
4_1	PME	8	5	3	1	2	62.5	62.5
	FW	**7**	**6**	**2**	**1**	**0**	**85.7**	**75.0**
	SWC	9	5	3	2	2	62.5	55.6
1_5	PME	20	11	7	3	6	55.0	61.1
	FW	**21**	**15**	**3**	**2**	**4**	**71.4**	**83.3**
	SWC	18	9	9	2	7	50.0	50.0
16_1	PME	14	8	6	4	2	57.1	57.1
	FW	**13**	**10**	**4**	**1**	**2**	**76.9**	**71.4**
	SWC	19	7	7	3	9	36.8	50.0
216_11	PME	18	4	6	4	10	22.2	36.7
	FW	**13**	**9**	**2**	**2**	**3**	**69.2**	**81.8**
	SWC	7	4	3	1	2	57.1	36.7

Accuracy rate representative of the correct amount is divided by the number extracted, recall rate representative of the correct amount is divided by the effective number of keyframes
PME is the algorithm of perceived motion energy, *SWC* is the algorithm based on correlation, *FW* is the algorithm of this chapter based on the feature value. Blackbody is the result of the experiment based on eigenvalues. Accuracy and recall are higher than the other two methods

References

1. Momin, B. F., & Rupnar, G. B. (2016). Keyframe extraction in surveillance video using correlation. In *2016 International Conference on Advanced Communication Control and Computing Technologies (ICACCCT)* (pp. 276–280). Piscataway: IEEE.
2. Xu, H., & He, Y. (2016). A video image preprocessing method for underground coal mine monitoring. *Chinese Journal of Industry and Mine Automation, 42*(1), 32–34.
3. Lin, Y.-C., & Lian, F.-L. (2014). Data reduction based on keyframe with motion energy extraction rules. In *Proceeding of the IEEE International Conference on Information and Automation Hailar, 2014* (pp. 507–512). Piscataway: IEEE.
4. Guan, G., Wang, Z., Lu, S., Da Deng, J., & Feng, D. D. (2013). Keypoint-based keyframe selection. *IEEE Transactions on Circuits and Systems for Video Technology, 23*(4), 729–734.
5. Sharma, C., & Sathish, P. K. (2015). Video content and structure description based on keyframes, clusters and storyboards. In *2015 IEEE 17th International Workshop on Multimedia Signal Processing (MMSP)* (pp. 245–249). Piscataway: IEEE.
6. Liu, Z., He, S., Hu, W., & Li, Z. (2017). Video sequence moving target detection based on background subtraction. *Chinese Journal of Computer Application, 37*(6), 1777–1781.

7. Jacques, J. C. S., Jr., Jung, C. R., & Musse, S. R. (2005). Background subtraction and shadow detection in grayscale video sequences. In *Proceedings of the XVIII Brazilian Symposium on Computer Graphics and Image Processing (SIBGRAPI'05)* (pp. 1530–1834/05). Piscataway: IEEE.
8. Feng, W., & Liu, B. (2017). Improved SIFT algorithm image matching research. *Chinese Journal of Computer Engineering and Application, 1*(1), 1–12.
9. Barbieri, T. T. d. S., & Goularte, R. (2014). KS-SIFT: A keyframe extraction method based on local features. In *2015 International Conference on Industrial Instrumentation and Control (ICIC)* (pp. 13–17). Piscataway: IEEE.

An Introduction to Formation Control of UAV with Vicon System

Yangguang Yu, Zhihong Liu, and Xiangke Wang

1 Introduction

In recent years, formation flight of UAV is widely applied in areas such as surveillance, aerial photography, and attacking. Despite its practical potential in various applications, the formation control, especially of a large swarm, still exists theoretical and engineering challenges in coordination and control of a large number of UAVs. Therefore, formation control of a large swarm receives considerable attentions and is intensively investigated. Fortunately, owing to the contributions of numerous researchers in the world, the formation control of a large swarm makes a major breakthrough and many splendid engineering works are demonstrated in the last year. The LOCUST Project directed by ONR of USA achieved rapid launch of 30 autonomous, swarming UAVs. On Nov. 6th, 2016, Intel made an impressive light show with a fleet of 500 drones. Just on Jan. 9th, 2017, Department of Defense in the USA announces successful micro-drone demonstration. In the demonstration, 103 Perdix drones were launched from three F/A-18 Super Hornets. The micro-drones demonstrated advanced swarm behaviors such as collective decision-making, adaptive formation flying, and self-healing.

In the research area, a significant amount of research efforts have been focused on formation control. According to the types of sensed and controlled variables, formation control can be categorized as position [1], displacement [2], and distance based [3]. Some advanced mathematical theories are introduced in the formation control, such as graph theory [4] and bearing rigidity [5]. While as many techniques have been used, the Vicon system is becoming popularly used in the control or

Y. Yu · Z. Liu · X. Wang (✉)
College of Mechatronics and Automation, National University of Defense Technology, Changsha, P. R. China
e-mail: xkwang@nudt.edu.cn

© Springer Nature Switzerland AG 2020
H. Lu, L. Yujie (eds.), *2nd EAI International Conference on Robotic Sensor Networks*, EAI/Springer Innovations in Communication and Computing, https://doi.org/10.1007/978-3-030-17763-8_17

evaluation of control performance of UAV recent years. Vicon system is a passive optical motion capture system which can record the movement of objects or people with high accuracy. Many famous universities or laboratories have introduced Vicon system into their experiment, such as University of Pennsylvania and University of Southern California. In [6], Alex Kushleyev et al. describe a micro quadrator with onboard attitude estimation and control that operates autonomously with Vicon system as external localization system. In [7], a 49-vehicle formation flight is demonstrated by using the motion capture localization. Vicon system is also utilized by other institutions to do researches and some references there in [8, 9].

In this paper, we outline the system framework for indoor experiment of multiple UAVs with Vicon system. Firstly, we illustrate the hardware setup for experiment as well as the architecture of them. Next, a distributed software architecture is described in detail. Then, we demonstrate a distributed flight demo in which three quadcopters circle around a fixed point coordinately by using the consensus protocol. Finally, the flight data of demo is recorded and analyzed.

2 System Framework

2.1 Hardware Setup

We outline our system in Fig. 1. Using the spherical marker attached to the quadcopter, the vehicles are tracked by Vicon motion capture system. Their position and attitude information are sent to their onboard processor via TCP/IP protocol

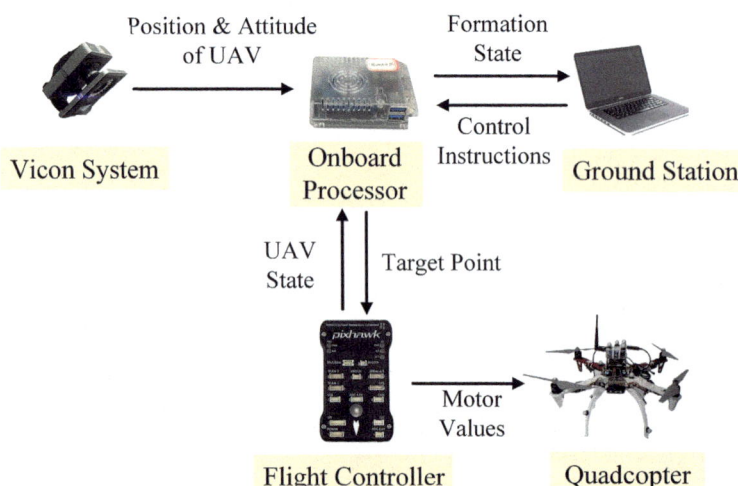

Fig. 1 The hardware architecture of system

with a frequency of 100 Hz. At the same time, the onboard processor receives the control instructions from ground station and UAV state from flight controller as well. Based on this information, the onboard processor calculates and forms the formation state and UAV's target point in the next step, which is received by ground station and flight controller, respectively. Finally, according to its target point, the flight controller calculates the motor values and send it to quadcopter. The whole information flow in the whole system is depicted in Fig. 2. The details of hardware component are described as follows.

The vehicle we use is a quadcopter similar to DJI FlameWheel 450. The diameter of quadcopter is approximately 45 cm and it is a suitable size for indoor flight experiment. The load capacity of our vehicle is approximately 2.5 kg, which makes it possible to equip quadcopter with other peripheral devices (like camera, etc.). As for endurance, the flight time of a single flight can be up to more than 10 min, which is enough for most of the experiments. Some peripheral devices are attached to the vehicle, depicted in Fig. 3.

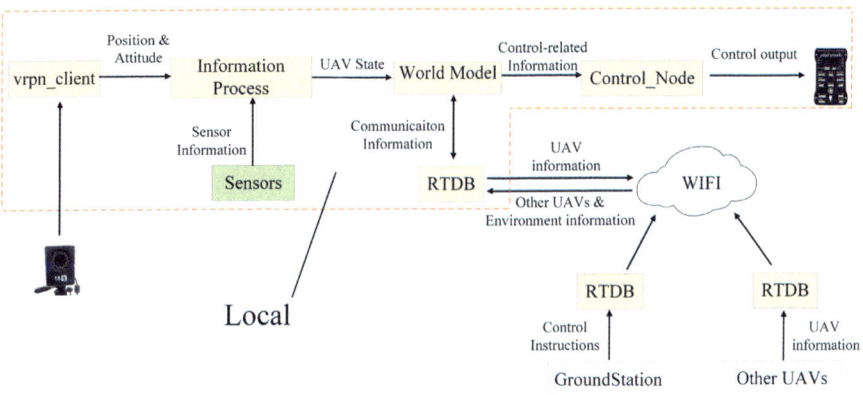

Fig. 2 The software architecture of the whole system

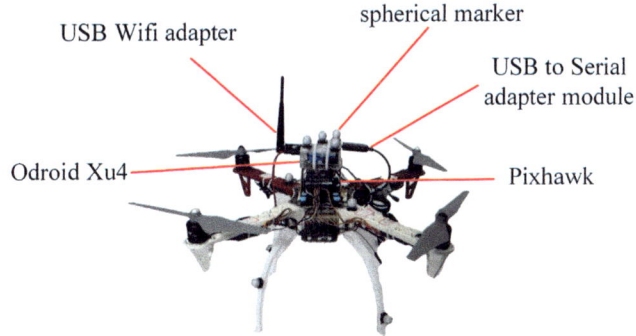

Fig. 3 The vehicle for experiment and its setup

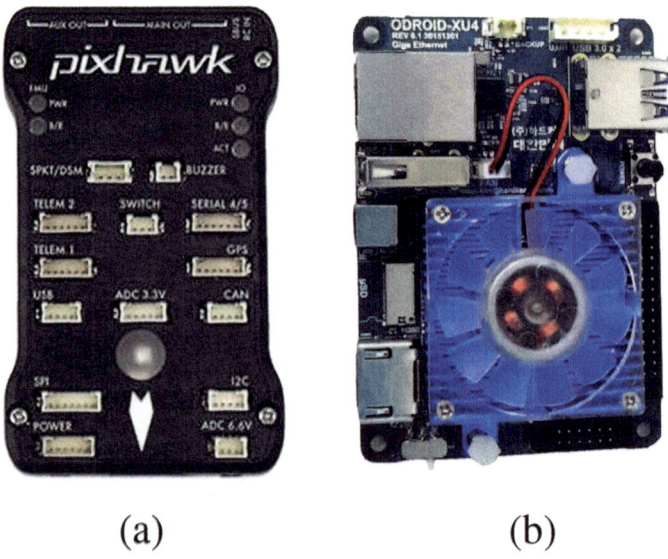

Fig. 4 Experiment hardware. (**a**) Pixhawk; (**b**) Odroid-Xu4

The flight controller we choose is the Pixhawk (depicted in Fig. 4a), which is produced by the famous open-hardware manufacturer 3DR. Pixhawk is an industry standard autopilot and has a stable and high-level calculation performance. In addition, Pixhawk owns a variety of input and output interfaces, which provides a great convenience for equipping quadcopter with other peripheral devices.

The onboard processor we choose is Odroid-XU4, which is a heterogeneous multi-processing (HMP) octa core linux computer (depicted in Fig. 4b). It is a new generation of computing device with more powerful, more energy-efficient hardware and a smaller form factor. By implementing the eMMC 5.0, USB 3.0, and gigabit ethernet interfaces, the Odroid-XU4 boasts amazing data transfer speeds, a feature that is increasingly required to support advanced processing power on ARM devices. It has two USB3.0 host, a USB 2.0 host, and a gigabit ethernet port utilized by a USB WiFi adapter. In addition, in order to achieve the communication between Odriod and Pixhawk, one of the Odroid's USB3.0 host and Pixhawk's serial port is connected by a USB to serial adapter module.

2.2 Software Architecture

The software architecture of the whole system is illustrated in Fig. 2. Basically, the software architecture is under ROS (robot operation system) framework. As the number of UAVs in a swarm is large, it is difficult to rewrite the control code for every UAV if some trouble occurs. Therefore, all the code for each UAV

Algorithm 1 Coordinated flight

1: Init target position
2: Calculate the distance D between the position of UAV and target point
3: **if** D ≤ threshold **then**
4:    Set the flag IsReached=true
5: **else**
6:    Set the flag IsReached=false
7: **if** IsReached==true **then**
8:    update the value of target angle as $\phi_i(k+1) = \sum_{j \in \mathscr{N}_i} \phi_j - 2\phi_i(k) + \triangle\phi$
9: **else**
10:    keep the value of target angle unchanged
11: Calculate the target position according to the target angle
12: Calculate the control output according to the position of UAV and target point

---

should be the same and the UAV is identified by an ID number. Mainly four nodes (colored as yellow) are included in the framework. The vrpn_client is an open ROS package developed by University of Toronto and can be found in https://github.com/clearpathrobotics/vrpn_client_ros. It serves as a driver providing data from VICON motion capture systems and publish the data in the form of ROS topic. Both of the information provided by the vrpn_client and sensors are sent to information node and fused. Specifically, here an extended Kalman filter is used to fuse the IMU data with the Vicon data and further the UAV can get its local position and other states. In World Model node, all the information of UAV and environment is integrated. The control node calculate the control output according to the control algorithm based on the related information from World Model node, including UAV states, environment situation, ground station constructions, etc. Finally, the control output is sent to Pixhawk via the serial interface.

In multi-UAV system, information need to be exchanged frequently among UAVs as well as between ground station and UAVs. Therefore, it is very important to ensure a stable data link in multi-UAV system. In analogy to many parallel programs in other institutions, the communication of our indoor flight demo is based on the WiFi, which is convenient and easily accessible. At the beginning, we tried to use the ROS protocol to achieve the communication among UAVs and ground station. However, network congestion occurs using this method. In order to solve this issue, a communication package named RTDB (real-time database) is introduced into our framework. In essence, RTDB is a communication protocol that adapts itself to the current conditions of the medium and attempts to deliver an adequate timeliness while using the least bandwidth [10]. Through RTDB protocol, the World Model node exchanges information with other UAVs and ground station. The physical experiment verifies that the system runs well under this framework.

3 Experimental Setup

3.1 Control Algorithm

To verify the effectiveness and feasibility of our system, a simple demo is designed. Three quadcopters are designed to fly around a fixed point coordinately and keep an equal distance from each other. To make the UAV circle around a fixed point, we use an online plan method based on position control. Firstly, 120 key points are extracted on the planned circle evenly. The target key point is sent to the controller of UAV successively when the UAV reaches a key point. Because all the key points are on the planned circle, their position p can be characterized by the angle between positive y-axis and the line connecting UAV's position and the origin. Thus, the trajectory of UAV can be planned based on the defined angle.

To make the quadcopters circle coordinately and keep an equal distance from each other, a simple consensus protocol is used. The angle defined previously is chosen as consensus state and a Laplacian matrix (see [4]) is defined as follows:

$$L = \begin{bmatrix} -2 & 1 & 1 \\ 1 & -2 & 1 \\ 1 & 1 & -2 \end{bmatrix} \tag{1}$$

Let $\phi = (\phi_1, \phi_2, \phi_3)^T \in \mathbb{R}^3$ denote the angle of three quadcopters. To make the quadcopters keep an equal distance from each other, the consensus protocol of system for each control step can be designed as $\phi(k+1) = L\phi(k)$. Further, in order to make quadcopters fly circle, an increment constant $\triangle\phi$ should be considered. Therefore, the target angle for each quadcopter is as follows:

$$\phi_i(k+1) = \sum_{j \in \mathcal{N}_i} \phi_j - 2\phi_i(k) + \triangle\phi \tag{2}$$

where $\mathcal{N}_i$ represents the other two quadcopters. The details of algorithm are described in Algorithm 1 and illustrated in Fig. 5.

3.2 Results and Analysis

We did our demo in a $20 \times 15 \times 5$ m laboratory, where a total of 24 Vicon cameras are utilized. The experiment environment and experiment scene are depicted in Figs. 6 and 7.

The trajectory of target position and true position is recorded and illustrated in Fig. 8. As revealed in Fig. 8, although the trajectory of quadcopter is a circle on the whole, some fluctuations around the set point still exist. The reason for these phenomena is multifold. On the one hand, as the circle radius is just 2 m, small

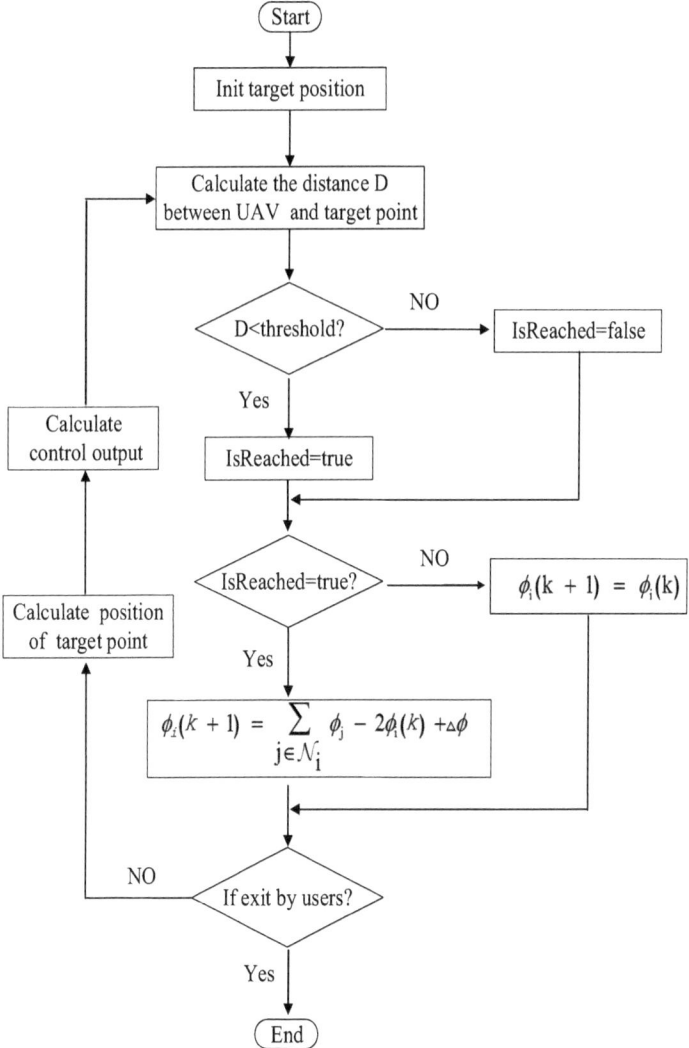

Fig. 5 The flow chart of Algorithm 1

fluctuations are still very obvious. One the other hand, the control algorithm also needs to be improved and some parameters need to be further adjusted.

In order to verify the coordination of quadcopters, the angle among quadcopters is calculated according to their position and illustrated in Fig. 9. The angle between any two quadcopters should be 120° (illustrated as black line). However, the experiment results are not very satisfactory. As analysis above, the experimental setup as well as control algorithm affects the experimental performance. In essence, the velocity control should be introduced into the control framework to obtain ideal performance. We will do this work in the next step.

Fig. 6 Experiment environment

Fig. 7 Experiment scene

4   Conclusion and Future Work

In this paper, we present a framework for multi-UAV control with aid of Vicon system. Both the hardware setup and software architecture are introduced in detail. Applying this framework in our experiment, we demonstrate a decentralized coordinated flight demo in which three quadcopters circle around a fixed point. In the end of paper, the flight data are reported and analyzed.

From the results of experiment, the controller of quadcopters needs to be greatly improved. In addition, more advanced algorithm can be applied in this framework. Therefore, future work will focus on the improvement for the controller

Fig. 8 The target trajectory
and the true trajectory of UAV

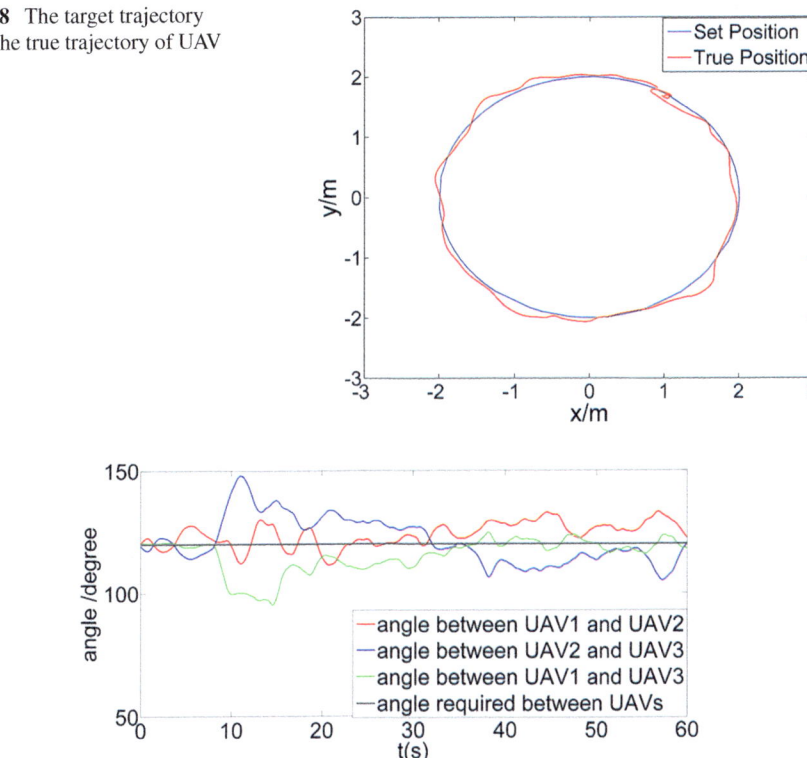

Fig. 9 The angle between UAVs

of the quadcopters. Moreover, more complicated application of multi-UAV should be achieved in this framework, such as encirclement and formation tracking. At present, the environment information the UAV receives is provided by the Vicon system. In our future work, the sensing technology will also be utilized to replace the Vicon system. Thus the UAV can achieve fully autonomous.

References

1. Ren, W., & Atkins, E. (2007). Distributed multi-vehicle coordinated control via local information exchange. *International Journal of Robust and Nonlinear Control, 17*(10–11), 1002–1033.
2. Coogan, S., & Arcak, M. (2012). Scaling the size of a formation using relative position feedback. *Automatica, 48*(10), 2677–2685.
3. Hendrickx, J. M. Anderson, B., Delvenne, J. -C., & Blondel, V. D. (2007). Directed graphs for the analysis of rigidity and persistence in autonomous agent systems," *International Journal of Robust and Nonlinear Control, 17*(10–11), 960–981.
4. Ren, W., Beard, R. W., & Atkins, E. M. (2007). Information consensus in multivehicle cooperative control. *IEEE Control Systems Magazine, 27*(2), 71–82.

5. Zhao, S., & Zelazo, D. (2016). Bearing rigidity and almost global bearing-only formation stabilization. *IEEE Transactions on Automatic Control, 61*(5), 1255–1268.
6. Kushleyev, A., Mellinger, D., Powers, C., & Kumar, V. (2013). Towards a swarm of agile micro quadrotors. *Autonomous Robots, 35*(4), 287–300.
7. Preiss, J. A., Hönig, W., Sukhatme, G. S., & Ayanian, N. (2017). Crazyswarm: A large nano-quadcopter swarm. In *2017 IEEE International Conference on Robotics and Automation (ICRA)*. Piscataway: IEEE.
8. Thomas, J., Polin, J., Sreenath, K., & Kumar, V. (2013). Avian-inspired grasping for quadrotor micro UAVs. In *ASME 2013 International Design Engineering Technical Conferences and Computers and Information in Engineering Conference*. New York, NY: American Society of Mechanical Engineers.
9. Grabe, V., Bülthoff, H. H., & Giordano, P. R. (2012). On-board velocity estimation and closed-loop control of a quadrotor UAV based on optical flow. In *2012 IEEE International Conference on Robotics and Automation (ICRA)* (pp. 491–497). Piscataway: IEEE.
10. Santos, F., Almeida, L., Pedreiras, P., Lopes, L. S., & Facchinetti, T. (2004). An adaptive TDMA protocol for soft real-time wireless communication among mobile autonomous agents. In *Proceedings of the International Workshop on Architecture for Cooperative Embedded Real-Time Systems, WACERTS*, vol. 2004 (pp. 657–665).

Quadratic Discriminant Analysis Metric Learning Based on Feature Augmentation for Person Re-Identification

Cailing Wang, Hao Qi, Guangwei Gao, and Xiaoyuan Jing

1 Introduction

Person re-identification is the technique to judge whether person under different cameras are the same pedestrian. Existing research methods for person re-identification can be roughly divided into two categories: (1) feature representation, design and extraction of the most robust features on persons to better represent persons; (2) subspace/metric learning, projecting the extracted features into a certain space, in which the same person is as close as possible, as far as possible between different persons scattered, and a robust distance is learned in the training set to solve complex matching problems. This paper focuses on subspace/metric learning methods. Subspace/metrics learning methods can be roughly divided into two categories: one is a general subspace metric for all cameras, learning a suitable projection matrix for all cameras. This method has strong generalization ability but weakens the discriminative ability of features within each camera; the other is specific subspace metric for specific camera, learning a projection matrix for each camera. This method extracts features with strong discriminative ability but poor generalization ability. This paper is to improve XQDA metric learning method, so that it has better generalization ability and discriminative ability at the same time.

C. Wang · H. Qi (✉) · G. Gao · X. Jing
Nanjing University of Posts & Telecommunications, Nanjing, China

© Springer Nature Switzerland AG 2020
H. Lu, L. Yujie (eds.), *2nd EAI International Conference on Robotic Sensor Networks*, EAI/Springer Innovations in Communication and Computing,
https://doi.org/10.1007/978-3-030-17763-8_18

2 Related Work

Person re-identification is a problem of finding a person from a gallery set who has the same identity to the probe. Its application is very wide [1, 2]: pedestrian tracking, pedestrian search, public security, etc. However, the technology still has many challenges [3], because the images captured by different cameras have great distortion, such as environment, lighting, person pose, camera properties or viewpoint, and occlusion which cause great interference for person re-identification.

Feature representation: Designing and extracting features that are more robust on the person appearances to represent person. Color features are widely used in person re-identification. However, color features are easily affected by external factors such as lighting, environment, and camera properties. Texture features are very robust to lighting and the environment, but they are sensitive to noise. For the feature representation, many effective methods have been proposed, such as SDALF [4], LDFV [5], salience matching [6], mid-level filters [7], and so on. But how to design and extract robust features is still a challenging problem.

Subspace/metric learning [8–13]: Subspace metric learning methods can be roughly classified into two categories: camera general projection matrix learning [14–17] and camera special projection matrix learning [18–22]. General projection matrix means that all cameras share a projection matrix, and the person image under each camera is transformed by the projection matrix and then measured. This method has strong generalization ability, but ignores the features variation in each camera, thereby reducing the discriminative ability of person features in each camera. The special projection matrix is to establish a projection matrix for each camera. Each camera performs an independent feature transformation. This method takes into account the internal relationship of each camera and has good discriminative ability for the features of each camera. However, its generalization ability is weak. In order to solve the above problems, this paper improves the general projection metric learning XQDA method, and increases its discriminative ability while focusing on generalization ability.

3 Approach

3.1 A Brief Introduction to XQDA

Person representation: $X = [x_1, x_2, \ldots, x_n] \in R^{d \times n}$, n is the number of samples in camera a, and d is the dimension of the feature; $Y = [y_1, y_2, \ldots, y_m] \in R^{d \times m}$, m is the number of samples in camera b. X denotes all the samples in camera a, x_i denotes ith person in camera a, and Y is similar to this. When $i = j$, the same person is under different cameras

$$\Delta = x_i - y_j \tag{1}$$

When $i = j$, Δ is called the inter-person difference; otherwise, it is called the extern-person difference. Ω_I and Ω_E are denoted inter-person variations and extern-person variations, respectively.

Under the zero-mean Gaussian distribution, the likelihoods of observing Δ in Ω_I and Ω_E are defined as:

$$P\left(\Delta|\Omega_I\right) = \frac{1}{(2\Pi)^{\frac{d}{2}}|\Sigma_I|^{\frac{1}{2}}} e^{\frac{-\Delta^T \Sigma_I^{-1}\Delta}{2}} \tag{2}$$

$$P\left(\Delta|\Omega_E\right) = \frac{1}{(2\Pi)^{\frac{d}{2}}|\Sigma_E|^{\frac{1}{2}}} e^{\frac{-\Delta^T \Sigma_E^{-1}\Delta}{2}} \tag{3}$$

where Σ_I and Σ_E are the covariance matrices of Ω_I and Ω_E, respectively.

By applying the Bayesian rule and the log-likelihood ratio test, the decision function can be simplified as:

$$f\left(\Delta\right) = \Delta^T \left(\Sigma_I^{-1} - \Sigma_E^{-1}\right) \Delta \tag{4}$$

and so the derived distance function between x_i and y_j is

$$d\left(x_i, y_j\right) = \left(x_i - y_j\right)^T W \left(\Sigma_I^{-1} - \Sigma_E^{-1}\right) W^T \left(x_i - y_j\right) \tag{5}$$

XQDA learns a projection matrix W projecting the raw data into a lower-dimensional subspace and learns a metric in this new space for re-identification. So, the distance function Eq. (5) in the new space is computed as:

$$d_W\left(x, y\right) = \left(x - y\right)^T W \left(\Sigma_I'^{-1} - \Sigma_E'^{-1}\right) W^T \left(x - y\right) \tag{6}$$

where $\Sigma_I' = W^T \Sigma_I W$ and $\Sigma_E' = W^T \Sigma_E W$.

However, directly optimizing Eq. (6) is difficult. So, by the derivation of the equivalent transformation, Eq. (6) is equivalent to:

$$J(W) = \frac{W^T \Sigma_E W}{W^T \Sigma_I W} \tag{7}$$

The maximization of Eq. (7) is equivalent to:

$$\max_W W^T \Sigma_E W, \ s.t. W^T \Sigma_I W = 1 \tag{8}$$

By solving Eq. (8), a projection matrix W can be obtained and then the metric matrix can be obtained.

The projection matrix W learned by this method is a general projection for all the cameras; therefore, the feature discriminability of this space is weak. This paper is to improve the method by improving the discriminability of the projection matrix, so as to achieve a better re-identification performance.

3.2 Feature Augmentation

The zero-padding augmentation of each camera can be rewritten as:

$$X_{aug} = \begin{bmatrix} I \\ 0 \end{bmatrix} X, \; Y_{aug} = \begin{bmatrix} 0 \\ I \end{bmatrix} Y \tag{9}$$

To generalize Eq. (9) as:

$$X_{aug} = \begin{bmatrix} R \\ M \end{bmatrix} X, \; Y_{aug} = \begin{bmatrix} M \\ R \end{bmatrix} Y \tag{10}$$

R and M definition as:

$$R = \frac{1+r}{Z} I, \; M = \frac{1-r}{Z} I \tag{11}$$

$$Z = \sqrt{(1-r)^2 + (1+r)^2}$$

By controlling the parameter, the corresponding R and M can be obtained.

3.3 The Proposed Approach

The mapping function of the general projection matrix of the camera can be expressed as:

$$f_a(X) = W^T X, \; f_b(Y) = W^T Y \tag{12}$$

Bring Eq. (10) into Eq. (12) as:

$$f_a\left(X_{aug}\right) = W^T X_{aug}, \; f_b\left(Y_{aug}\right) = W^T Y_{aug} \tag{13}$$

Decompose W into two parts as:

$$W = \left[W_1^T, W_2^T \right]^T \tag{14}$$

Bring Eq. (14) into Eq. (13) as:

$$\begin{aligned} f_a \left(X_{\text{aug}} \right) &= \left[W_1^T, W_2^T \right] X_{\text{aug}} = \left(W_1^T R + W_2^T M \right) X \\ f_b \left(Y_{\text{aug}} \right) &= \left[W_1^T, W_2^T \right] Y_{\text{aug}} = \left(W_1^T M + W_2^T R \right) Y \end{aligned} \tag{15}$$

It can be seen from Eq. (15) that the discriminative ability of XQDA can be increased through feature augmentation.

Let $||W_1 - W_2||^2$ be the regularization, and combining with the common ridge regularization as:

$$\begin{aligned} & \| W_1 - W_2 \|^2 + \eta \text{tr} \left(W^T W \right) = \text{tr} \left(W^T \begin{bmatrix} I & -I \\ -I & I \end{bmatrix} W \right) + \eta \text{tr} \left(W^T W \right) \\ & = (1 + \eta) \, \text{tr} \left(W^T C W \right) \end{aligned} \tag{16}$$

where $\eta < 0$

$$C = \begin{bmatrix} I & -\beta I \\ -\beta I & I \end{bmatrix}, \beta = \frac{1}{1+\eta} < 1$$

The above r and β can be obtained by the principle angles between the cameras. G_a and G_b are samples of camera a and camera b, respectively. The principle angles can be computed by singular value decomposition of $G_a^T G_b$ [23]:

$$G_a^T G_b = V_1 \cos (\Theta) V_2^T \tag{17}$$

where $\cos(\Theta) = \text{diag} (\cos(\theta_1), \cos(\theta_2), \ldots, \cos(\theta_d))$ and θ_i are the principle angles.

With the principle angles, r and β can be set as:

$$r = \frac{1}{d} \sum_{k=1}^d \left(1 - \cos^2 (\theta_k) \right), \quad \beta = \frac{1}{d} \sum_{k=1}^d (1 - \cos (\theta_k)) \tag{18}$$

Equation (16) is used as the regularization term of this paper to optimize the XQDA, so that its features have better discriminability. The final objective function can be expressed as:

$$\max_{W} J \left(W^T X_{\text{aug}} \right) + \lambda \text{tr} \left(W^T C W \right), s.t. g \left(W^T X_{\text{aug}} \right) \tag{19}$$

where J is the objective function and g are the constraints, and λ is an optimized parameter that can be predefined.

Since $\beta < 1$, $C = P \Lambda P^T$, and thus $\Lambda = P^T C P$. Let W be:

$$W = P \Lambda^{\frac{1}{2}} H \tag{20}$$

Thus $W^T C W = H^T H$, Eq. (19) can be equally defined as:

$$\max_{W} J \left(H^T \Lambda^{-\frac{1}{2}} P^T X_{\text{aug}} \right) + \lambda tr \left(H^T H \right), s.t.g \left(H^T \Lambda^{-\frac{1}{2}} P^T X_{\text{aug}} \right) \tag{21}$$

Get H and bring H into Eq. (20) to get the final projection matrix, and projecting the features of augmentation into a new space through the final projection matrix. On the basis of maintaining the generalization ability, the discriminative ability has been greatly improved, and the XQDA performance of re-identification has been further improved.

4 Experiment

In this experiment, we used VIPeR dataset, and 632 persons were randomly divided into two non-overlapped parts. Three hundred and sixteen persons were used as training sets to train the projection matrix, and 316 persons were used as test sets to test the performance of the method proposed in this paper. λ is predefined as 0.5. The features used in the experiments are the features provided by the KCCA [20] method. For the fairness of the experiments, all the methods used for comparison in this paper use this kind of features. CMC refers to the cumulative matching characteristic curve, which is a commonly used evaluation method in person re-identification (see Fig. 1). The rank-k recognition rate indicates the probability of correctly identifying person in the first K images. In order to reduce the experimental error, the experiment was repeated ten times and the average value was taken as the final experimental result.

Compared with the previous method, the experimental results are shown in Fig. 1 and Table 1. As can be clearly seen from Table 1, the performance of the method in this paper improves 3.95% than XQDA [16] at rank-1, and 6.08% at rank-5, and at rank-10, rank-15, and rank-20, the performance increased 5.03%, 4.12%, and 2.91%, respectively. Compared with the recently proposed MKSSL [13] and CSRML [11] algorithms, the performance has also improved at different rank levels. The reason is that the method in this paper improves feature discriminability in each camera by feature augmentation while maintaining its good generalization capability, thereby improving the final re-identification performance.

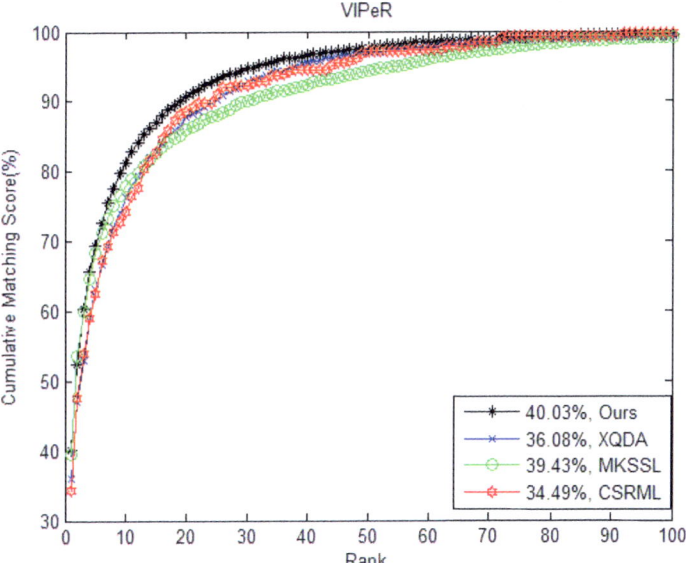

Fig. 1 Various methods of CMC on the VIPeR dataset

Table 1 The correct recognition rate on the VIPeR dataset (%)

	VIPeR				
Rank	1	5	10	15	20
Ours	40.03	69.37	81.30	87.09	90.63
XQDA [16]	36.08	63.29	76.27	82.97	87.72
MKSSL [13]	39.43	68.41	78.04	82.66	85.92
CSRML [11]	34.49	62.66	74.37	82.91	88.61

5   Conclusion

In this paper, the feature augmentation method is used to improve the problem of poor discriminability of XQDA for each camera. The current measure method is either to learn a general projection matrix for all cameras or to learn special projection matrix for each camera. The former generalization ability is better, but it loses the inherent properties of each camera, making the discriminative ability in each camera weaker; later the camera internal discriminative ability is focused on, but it does not well consider the relationship between cameras, and as a result the generalization ability is not very good. Based on the XQDA method, this paper considers the relationship between the inside and the outside of each camera at the same time to ensure its generalization ability and improve its discriminative performance. Finally, experiments are carried out on a widely used datasets, and the results show that the proposed method significantly improves the performance of the re-identification.

Acknowledgements This work was partially supported by the National Natural Science Foundation of China under Grant no. 61402237, 61502245, 61772568 and the Natural Science Foundation of Jiangsu Province under Grant no. BK20150849. Guangwei Gao is the corresponding author.

References

1. Serikawa, S., & Lu, H. (2014). *Underwater image dehazing using joint trilateral filter*. Oxford: Pergamon Press.
2. Lu, H., Li, Y., Uemura, T., et al. (2018). Low illumination underwater light field images reconstruction using deep convolutional neural networks. *Future Generation Computer Systems, 82*, 142–148.
3. Xu, X., He, L., Lu, H., et al. (2019). Deep adversarial metric learning for cross-modal retrieval. *World Wide Web, 22*(2), 657–672.
4. Bazzani, L., Cristani, M., & Murino, V. (2013). Symmetry-driven accumulation of local features for human characterization and re-identification. *Computer Vision & Image Understanding, 117*(2), 130–144.
5. Ma, B., & Su, Y. (2012). Local descriptors encoded by fisher vectors for person re-identification. In *International conference on computer vision* (pp. 413–422). Berlin: Springer.
6. Zhao, R., Ouyang, W., & Wang, X. (2014). Person re-identification by salience matching. In *IEEE International Conference on Computer Vision* (pp. 2528–2535).
7. Zhao, R., Ouyang, W., & Wang, X. (2014). Learning mid-level filters for person re-identification. In *IEEE Conference on Computer Vision and Pattern Recognition* (pp. 144–151). IEEE.
8. Liao, S., & Li, S. Z. (2015). Efficient PSD constrained asymmetric metric learning for person re-identification. In *IEEE International Conference on Computer Vision* (pp. 3685–3693). IEEE.
9. Zhang, L., Xiang, T., & Gong, S. (2016). Learning a discriminative null space for person re-identification. In *IEEE Conference on Computer Vision and Pattern Recognition* (pp. 1239–1248). IEEE.
10. Chen, D., Yuan, Z., Chen, B., et al. (2016). Similarity learning with spatial constraints for person re-identification. In *IEEE Conference on Computer Vision and Pattern Recognition* (pp. 1268–1277). IEEE.
11. Wang, J., Zhu, J., Wang, Z., et al. (2016). Contextual similarity regularized metric learning for person re-identification. In *International Conference on Pattern Recognition* (pp. 2048–2053).
12. Kodirov, E., Xiang, T., & Gong, S. (2015). Dictionary learning with iterative Laplacian regularization for unsupervised person re-identification. In *British Machine Vision Conference* (pp. 44.1–44.12).
13. Yang, X., Wang, M., Hong, R., et al. (2017). Enhancing person re-identification in a self-trained subspace. *ACM Transactions on Multimedia Computing Communications & Applications, 13*(3), 27.
14. Hirzer, M. (2012). Large scale metric learning from equivalence constraints. In *IEEE Conference on Computer Vision and Pattern Recognition* (pp. 2288–2295). IEEE Computer Society.
15. Li, Z., Chang, S., Liang, F., et al. Learning locally-adaptive decision functions for person verification. In *Computer Vision and Pattern Recognition* (pp. 3610–3617). IEEE.
16. Liao, S., Hu, Y., Zhu, X., et al. (2015). Person re-identification by local maximal occurrence representation and metric learning. In *Computer Vision and Pattern Recognition* (pp. 2197–2206). IEEE.
17. Pedagadi, S., Orwell, J., Velastin, S., et al. (2013). Local fisher discriminant analysis for pedestrian re-identification. In *Computer Vision and Pattern Recognition* (pp. 3318–3325). IEEE.

18. An, L., Yang, S., & Bhanu, B. (2015). Person re-identification by robust canonical correlation analysis. *IEEE Signal Processing Letters, 22*(8), 1103–1107.
19. An, L., Kafai, M., Yang, S., et al. (2016). Person re-identification with reference descriptor. *IEEE Transactions on Circuits & Systems for Video Technology, 26*(4), 776–787.
20. Lisanti, G., Masi, I., & Bimbo, A. D. (2014). Matching people across camera views using kernel canonical correlation analysis. In *Proceedings of the International Conference on Distributed Smart Cameras* (pp. 1–6). ACM.
21. Chen, Y. C., Zheng, W. S., Lai, J. H., et al. (2017). An asymmetric distance model for cross-view feature mapping in person re-identification. *IEEE Transactions on Circuits & Systems for Video Technology, 27*(8), 1661–1675.
22. Chen, Y. C., Zheng, W. S., Lai, J. (2015). Mirror representation for modeling view-specific transform in person re identification. In *International Conference on Artificial Intelligence* (pp. 3402–3408). AAAI Press.
23. Hamm, J., & Lee, D. D. (2008). Grassmann discriminant analysis: a unifying view on subspace-based learning. In *Proceedings of the 25th International Conference on Machine Learning* (pp. 376–383). ACM.

Weighted Linear Multiple Kernel Learning for Saliency Detection

Quan Zhou, Jinwen Wu, Yawen Fan, Suofei Zhang, Xiaofu Wu, Baoyu Zheng, Xin Jin, Huimin Lu, and Longin Jan Latecki

1 Introduction

The human visual system (HVS) has an outstanding ability to quickly locate the most interesting parts in a given scene. Such image parts are considered as salient since it is assumed these parts attract greater attention than other parts by the

Q. Zhou (✉)
National Engineering Research Center of Communications and Networking, Nanjing University of Posts and Telecommunications, Nanjing, P. R. China

State Key Laboratory for Novel Software Technology, Nanjing University, Nanjing, P. R. China
e-mail: quan.zhou@njupt.edu.cn

J. Wu
College of Information Engineering, China University of Geosciences, Wuhan, P. R. China

S. Zhang
School of Internet of Things, Nanjing University of Posts and Telecommunications, Nanjing, P. R. China

Y. Fan · X. Wu · B. Zheng
National Engineering Research Center of Communications and Networking, Nanjing University of Posts and Telecommunications, Nanjing, P. R. China

X. Jin
Department of Cyber Security, Beijing Electronic Science and Technology Institute, Beijing, China

CETC Big Data Research Institute Co., Ltd., Guiyang, Guizhou, China

H. Lu
Department of Mechanical and Control Engineering, Kyushu Institute of Technology, Kitakyushu, Japan

L. J. Latecki
Department of Computer and Information Sciences, Temple University, Philadelphia, PA, USA

© Springer Nature Switzerland AG 2020
H. Lu, L. Yujie (eds.), *2nd EAI International Conference on Robotic Sensor Networks*, EAI/Springer Innovations in Communication and Computing,
https://doi.org/10.1007/978-3-030-17763-8_19

HVS. The recent study of saliency approaches may reveal the attention mechanisms of visual biology to predict human fixation selection behavior. Saliency detection plays a significant role in the fields of computer vision, and is involved in many visual applications, such as automatic image cropping [5], image thumbnailing [19], image/video compression [21], image segmentation [15], image quality assessment [17], and object detection/recognition [2].

The recent years have witnessed great progress in saliency detection, and it has received extensive attention by the researcher in the fields of psychologists and computer vision [5, 6, 12, 26, 28, 31]. As a pioneer work, Treisman [24] proposed a feature integration theory (FIT) which is composed by three main steps for HVS: (1) the bottom-up contrast computation based on simple low-level image stimuli signals, such as luminance, color, texture, and orientation, which are driven from the input image [12]; (2) the integration process via fusing various bottom-up feature maps produced in the first step[8, 11]; (3) the enhanced highlighting salient parts with the assistance of top-down priors if available[3, 22]. In spite of achieving promising results, these approaches are still suffered from the following limitations: The existing methods for feature map integration, such as average operation [12], selective fusing operation [8], max or min operation [30], are not flexible enough and adaptive sufficiently. They are not able to assign adaptive weights to predict visual saliency, which reflect the confidence level of each individual feature map.

This paper attempts to solve this problem using weighted linear multiple kernel learning (WLMKL) framework for the task of saliency detection. More specifically, our method firstly utilizes corner-surround contrast (CSC) [29] to measure the saliency for each pixel. Except computing the appearance contrast, this contrast operation also considers the relative location between center and surrounding regions, which enables us to predict more exact location of the salient parts. Thereafter, two types of commonly used contrast measurement, named CESC [12, 30] and global contrast (GC) [5], are calculated as complementary feature maps. Finally, to further investigate the contribution of each feature map, a multi-cues integration framework is designed using our WLMKL scheme to predict visual saliency. To optimize our WLMKL model, we design an EM-like procedure to alternatively update the model parameters and combined feature weights, where a closed-form solution can be obtained for updating feature weights. In summary, the main contributions of this paper are mainly summarized as follows:

- We propose to use WLMKL paradigm to formulate visual saliency, motivated by assigning adaptive weights to integrate different feature maps. Due to the duality, an efficient algorithm is designed to solve our WLMKL problem with ℓ_2-norm regularization. The proposed model benefits from the advantages of each feature map, while keeping the assigned weights is always normalized.
- We evaluate our approach for the tasks of visual saliency detection. We compared our model with the mainstream models in terms of detection accuracy. The experimental results show that our method outperforms these top-ranked models where previous studies have shown to be significantly predictive of salient parts in natural scenes.

This paper is organized as follows. We first elaborate on the detail of the proposed visual saliency detection method in Sect. 2. Section 3 reports the experimental results, and the conclusive remarks are given in Sect. 4.

2 The Proposed Method

The diagram of our approach is shown in Fig. 1. The training and testing images undergo the same procedure of feature map computation and combination. We first introduce the image representation using sparse coding technique, and then present the aforementioned contrast calculation and WLMKL framework.

2.1 *Image Representation*

From the perspective of human vision, a vision system should be adapted to the visual environment. As a supporting evidence for this theory, it has been shown that some neurons in V1 cortex resemble receptive fields that are learned via sparse coding algorithms [20]. From the perspective of computer vision, natural images are always with redundant structure, and thus can be sparsely represented by a set of localized and oriented filters [23]. To this end, we employ sparse coding technique to represent image patches, which has been demonstrated as an effective tool for saliency detection task [4, 10].

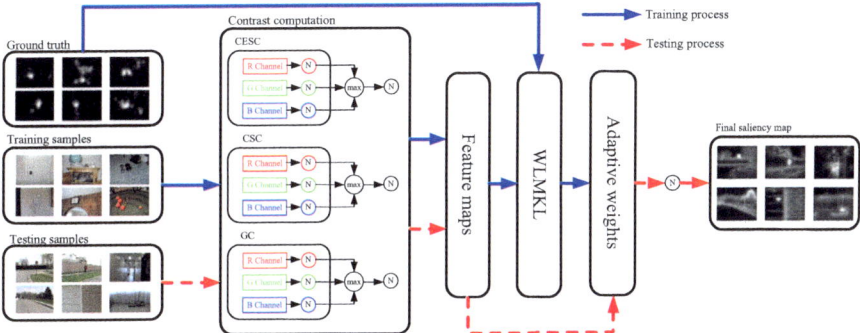

Fig. 1 Diagram of our saliency detection approach. In each channel of RGB color space, the CSC and CESC feature maps, which are the dissimilarity between a patch and its surroundings, and global feature map, which is based on rarity of an image patch with respect to the entire scene, are first computed and normalized. Then, the output feature maps and corresponding ground truth are used to train our WLMKL model. Finally, the feature maps of testing image are fed into the trained WLMKL model to generate final saliency map. The blue arrow denotes the training process and the dash red arrow indicates the testing process. (Best viewed in color)

Previously, the input image is first resized to $2^9 \times 2^9$ pixels. Suppose we have a series of image patches from the top-left to bottom-right of image with no overlap. Then the reconstructive coefficients $\boldsymbol{\eta}_i$ are calculated to represent patch $\mathbf{p}_i$ using the sparse coding algorithms [20]. In our implementation, we extracted 500,000 8×8 image patches randomly selected from color images using CIELab color space. Thus each basis in the dictionary is an $8 \times 8 = 64D$ vector, and we learn 200 dictionary functions. The sparse coding coefficients $\boldsymbol{\eta}_i$ are computed with the above learned basis using the LARS algorithm [18] implemented in the SPAMS toolbox.[1] Immediately below, we elaborate on the details of each blocks in Fig. 1.

2.2 Computing Contrast Feature Maps

CESC Inspired from the well-established computational architecture of [12, 30], the CESC operation, denoted as $f_{ce}^c(\mathbf{p}_i)$ in our model, is defined as the average weighted dissimilarity between a center patch $\mathbf{p}_i$ and its surrounding M neighborhood patches:

$$f_{ce}^c(\mathbf{p}_i) = \frac{1}{M} \sum_{m=1}^{M} \mathbf{W}_{im}^{-1} \mathbf{B}_{im} \tag{1}$$

where $\mathbf{W}_{im}$ is the Euclidean distance between the location of center patch $\mathbf{p}_i$ and the surround patch $\mathbf{p}_m$. $\mathbf{B}_{im}$ denotes the Euclidean distance between $\boldsymbol{\eta}_i$ and $\boldsymbol{\eta}_m$ in the feature space, vectors of coefficients for $\mathbf{p}_i$ and $\mathbf{p}_m$, respectively, where the Euclidean distance (ℓ_2 distance) is employed as distance measure. Superscript c denotes subchannels in RGB color space.

CSC It often happens that CESC may assign high saliency value to background, leading to incorrect detections. In order to overcome this shortcoming, we employ CSC operation $f_c^c(\mathbf{p}_i)$ to estimate visual saliency not only investigating the appearance difference but also relative location between center patch and its surrounding neighborhoods [29]. According to [29], four types of local contrast, namely bottom-right, bottom-left, top-right, and top-left templates, are defined. Let $f_{br}^c(\mathbf{p}_i)$, $f_{bl}^c(\mathbf{p}_i)$, $f_{tr}^c(\mathbf{p}_i)$, and $f_{tl}^c(\mathbf{p}_i)$ denote these four types of local contrast, respectively, the final CSC feature map is calculated as

$$f_c^c(\mathbf{p}_i) = f_{br}^c(\mathbf{p}_i) \times f_{bl}^c(\mathbf{p}_i) \times f_{tr}^c(\mathbf{p}_i) \times f_{tl}^c(\mathbf{p}_i) \tag{2}$$

For one specific type of local contrast (e.g., $f_{br}^c(\mathbf{p}_i)$), we define it as well as CESC computation that encodes the discriminative difference between corner patch

---

[1]http://www.di.ens.fr/willow/SPAMS/index.html.

$\mathbf{p}_i$ and its surrounding region. The same operation then applies to $f_{bl}^c(\mathbf{p}_i)$, $f_{tr}^c(\mathbf{p}_i)$, and $f_{tl}^c(\mathbf{p}_i)$, separately.

From Eq. (2), it is evidence that CSC assigns high value to patch $\mathbf{p}_i$ only when it is recommended by four types of local contrast simultaneously. Thus CSC is a more strict contrast operation than CESC, resulting in more effective to exclude outliers and inhibit background.

GC Sometimes, only using the local contrast operation may suppress areas within a homogeneous region, resulting in the uniformly highlighted salient regions and overemphasized object boundaries. Although the appearance cues of local patch may be similar to its neighbors, they are globally rareness with respect to the entire scene. To this end, we construct our global contrast feature map $f_g^c(\mathbf{p}_i)$ guided from the information-theoretic measure [4]. Instead of computing pixel saliency, here we calculate the inverse of probability $p(\mathbf{p}_i)$ for each patch over the entire scene as the global feature map:

$$f_g^c(\mathbf{p}_i) \propto - \sum_{j=1}^{n} \log(p(\eta_{ij})) \tag{3}$$

where η_{ij} is the jth component of vector $\boldsymbol{\eta}_i$. The GC assumes that coefficients η_{ij} are conditionally independent from each other, which is to some extent guaranteed by the sparse coding algorithm [20]. To construct the probability density function $(p(\eta_{ij}))$, we first calculate histogram distribution with B bins for each component η_{ij} among all image patches in the scene, then the distribution is normalized by dividing to its sum. If a patch is rare in one of the features, the product in From Eq. (3) will get a small value leading to high global contrast value for that patch overall.

2.3 WLMKL Framework

In this section, we design a WLMKL framework to estimate visual saliency, where the model parameters and adaptive feature weights are learned simultaneously. From the perspective of WLMKL, the feature maps are explicitly encoded through a set of so-called basic kernel functions $\{k_m\}_{m=1}^M$, then a SVM objective is employed to optimal model parameters and kernel feature weights.

Saliency Formulation Considering a training set Ω containing N samples, where each sample is characterized by M kinds of feature descriptors. Let $\Omega = \{(\mathbf{F}_n(\mathbf{x}), y_n \in \pm 1)\}_{n=1}^N$, $\mathbf{F}_n(\mathbf{x}) = \{f_{n,m}(\mathbf{x})\}_{m=1}^M$, where $f_{n,m}(\mathbf{x})$ is the feature map value for pixel $\mathbf{x}$, y_n is the target label for pattern $\mathbf{F}_n(\mathbf{x})$, where $+1$ denotes the pixel $\mathbf{x}$ is salient, and -1 indicates not. We use $\mathbf{F}_n$ to represent $\mathbf{F}_n(\mathbf{x})$ and $f_{n,m}$ to represent $f_{n,m}(\mathbf{x})$ for notation simplicity, then the saliency formulation for pixel $\mathbf{x}$ is defined as follows:

$$s(\mathbf{x}) = \sum_{n=1}^{N} \alpha_n y_n \mathbf{K}(\mathbf{F}_n, \mathbf{F}) + b \tag{4}$$

where α_n and b are model coefficients required to be learned from training samples, while $\mathbf{K}(\cdot, \cdot)$ represents a given ensemble kernel functions which are symmetric and positive definite. Our formulation considers that the kernel function $\mathbf{K}(\mathbf{F}_n, \mathbf{F})$ is a convex combination of basic kernels:

$$\mathbf{K}(\mathbf{F}_n, \mathbf{F}) = \sum_{m=1}^{M} \beta_m k_m(\mathbf{F}_n, \mathbf{F})$$

$$\beta_m \geq 0, \qquad \sum_{m=1}^{M} \beta_m = 1 \tag{5}$$

where $M = 3$ since we employ three different kinds of feature maps (CESC, CSC, and GC) to predict visual saliency, and β_m is the feature weight for the mth feature map. Keeping in mind that different feature maps may have different proportion of contribution for final saliency map, we thus constrain the feature weights in From Eq. (5) are non-negative and always normalized. Substitute From Eq. (5) to From Eq. (4), we get our final discriminative saliency model as:

$$s(\mathbf{x}) = \sum_{n=1}^{N} \alpha_n y_n \sum_{m=1}^{M} \beta_m k_m(\mathbf{F}_n, \mathbf{F}) + b$$

$$\beta_m \geq 0, \qquad \sum_{m=1}^{M} \beta_m = 1 \tag{6}$$

WLMKL Primal Learning Problem Actually, Eq. (6) corresponds to a standard support vector machine (SVM) formulation under MKL framework [7]. In order to identify the model parameters, we thus propose to address the following convex optimal problem, which we refer to as our primal WLMKL problem:

$$\min_{b, \, \xi, \, \beta, \, \mathbf{K}} \quad \frac{1}{2} \sum_{m} \frac{1}{\beta_m} ||k_m||^2 + C \sum_{n} \xi_n$$

$$s.t. \quad y_n[\sum_{m} k_m(\mathbf{F}_n) + b] \geq 1 - \xi_n, \forall n$$

$$\xi_n \geq 0, \forall n \tag{7}$$

$$\beta_m \geq 0, \qquad \sum_{m} \beta_m = 1$$

where $\boldsymbol{\beta} = \{\beta_m\}_{m=1}^M$, $\boldsymbol{\xi} = \{\xi_n\}_{n=1}^N$ are slack variables, and C is the trade-off parameter between training error and margin. It is clear that Eq. (7) is a primal learning problem involved in a weighted ℓ_2-norm regularization, where β_m controls the shape of the objective function.

Since this primal formulation is convex and differentiable, it provides a simple derivation of the dual problem [25]. By simply setting zero to the derivatives of the Lagrangian function for Eq. (7) with respect to the primal variables, we derive the associated dual problem as follows:

$$\max_{\boldsymbol{\alpha}} \min_{\boldsymbol{\beta}} J(\boldsymbol{\alpha}, \boldsymbol{\beta}) = -\frac{1}{2} \sum_{n,n'} \alpha_n \alpha_{n'} y_n y_{n'} \sum_m \beta_m k_m(\mathbf{F}_n, \mathbf{F}_{n'}) + \sum_n \alpha_n$$

$$s.t. \quad \sum_n \alpha_n y_n = 0 \quad 0 \le \alpha_n \le C \quad \forall n \tag{8}$$

$$\beta_m \ge 0, \quad \sum_m \beta_m = 1$$

where $\boldsymbol{\alpha} = \{\alpha_n\}_{n=1}^N$. Optimizing the coefficients $\boldsymbol{\alpha}$ and $\boldsymbol{\beta}$ is one particular form of the proposed WLMKL problems. Our approach utilizes such optimization to yield more flexible feature integration for visual saliency estimation.

Optimization Directly optimizing Eq. (8) is difficult, we thus resort to an iterative, EM-like strategy to alternately optimize $\boldsymbol{\alpha}$ and $\boldsymbol{\beta}$, separately. In each iteration, one of $\boldsymbol{\alpha}$ and $\boldsymbol{\beta}$ is optimized while the other is fixed, and then the roles of $\boldsymbol{\alpha}$ and $\boldsymbol{\beta}$ are switched. The whole iterations are repeated until convergence is reached.

On Optimizing $\boldsymbol{\alpha}$ Suppose we are given the optimized parameter $\boldsymbol{\beta}^*$, the optimization problem of Eq. (8) becomes

$$\max_{\boldsymbol{\alpha}} J(\boldsymbol{\alpha}) = -\frac{1}{2} \sum_{n,n'} \alpha_n \alpha_{n'} y_n y_{n'} \sum_m \beta_m^* k_m(\mathbf{F}_n, \mathbf{F}_{n'}) + \sum_n \alpha_n$$

$$s.t. \quad \sum_n \alpha_n y_n = 0 \quad 0 \le \alpha_n \le C \quad \forall n \tag{9}$$

which is identified as the standard SVM dual formulation using the combined kernel $\mathbf{K}(\mathbf{F}_n, \mathbf{F}) = \sum_m \beta_m^* k_m(\mathbf{F}_n, \mathbf{F})$. Thus the objective value $J(\boldsymbol{\alpha})$ can be obtained by any SVM algorithm.

On Optimizing $\boldsymbol{\beta}$ Suppose we are given the optimized parameter $\boldsymbol{\alpha}^*$, the optimization problem of Eq. (8) becomes

---

Algorithm 1: The training procedure of our algorithm

Input: Training data: $\mathbf{F}_1, \mathbf{F}_2, \cdots, \mathbf{F}_N$; Associated data label:
    $y_1, y_2, \cdots, y_N \in \{+1, -1\}$;
Initial kernel weights: $\boldsymbol{\beta} = \{\beta_1, \beta_2, \cdots, \beta_M\}$, where $\beta_m = \frac{1}{M}, m = \{1, \cdots, M\}$; Initial
temp weights: $T = 0$;
Basic kernel: Gaussian kernel; Step size: γ;
Stopping parameters: ε;
Output: Model coefficients: $\boldsymbol{\alpha}$; Basic kernel weights (feature map weights): $\boldsymbol{\beta}$

1 **for** $||\mathbf{T} - \boldsymbol{\beta}||_2 \geq \varepsilon$ **do**
2   Save current $\boldsymbol{\beta}$ as $T = \boldsymbol{\beta}$;
3   E-step: Optimize $\boldsymbol{\alpha}^*$
4   Compute $\boldsymbol{\alpha}^*$ using a standard SVM solver with fixed $\boldsymbol{\beta}$ and
         $k(\mathbf{F}_n, \mathbf{F}) = \sum_m \beta_m k_m(\mathbf{F}_n, \mathbf{F})$;
5   M-step: Optimize $\boldsymbol{\beta}^*$
6   Compute descent direction ∇J for $\boldsymbol{\beta}$ using Eq. (11);
7   Update $\boldsymbol{\beta}^*$ as $\boldsymbol{\beta}^* \leftarrow \boldsymbol{\beta} + \gamma \nabla J$;
8   Normalize $\boldsymbol{\beta}^*$ to satisfy the equality constraint in Eq. (10);
9 **end**
10 **return** $\boldsymbol{\alpha}$ and $\boldsymbol{\beta}$;

$$\min_{\boldsymbol{\beta}} J(\boldsymbol{\beta}) = -\frac{1}{2} \sum_{n,n'} \alpha_n^* \alpha_{n'}^* y_n y_{n'} \sum_m \beta_m k_m(\mathbf{F}_n, \mathbf{F}_{n'}) + \sum_n \alpha_n^*$$

$$\text{s.t.} \quad \sum_n \alpha_n^* y_n = 0 \quad 0 \leq \alpha_n^* \leq C \quad \forall n \tag{10}$$

$$\beta_m \geq 0, \qquad \sum_m \beta_m = 1$$

which is actually a non-linear objective function with constraints over the simplex. With our positivity definition on the kernel functions, $J(\boldsymbol{\beta})$ is convex and differentiable. Thus we solve this problem using a reduced gradient method. By simple differentiation of the objective function of Eq. (10) with respect to β_m, we have

$$\nabla J = \frac{\partial J(\boldsymbol{\beta})}{\partial \beta_m} = -\frac{1}{2} \sum_{n,n'} \alpha_n^* \alpha_{n'}^* y_n y_{n'} k_m(\mathbf{F}_n, \mathbf{F}_{n'}) \quad \forall n \tag{11}$$

Once the gradient of $J(\boldsymbol{\beta})$ is computed, $\boldsymbol{\beta}$ is updated using a descent direction ∇J as $\boldsymbol{\beta} \leftarrow \boldsymbol{\beta} + \gamma \nabla J$, where γ is the step size. Recalling from Eq. (10), the non-negative and normalized constraint is also required to be satisfied after $\boldsymbol{\beta}$ is updated.

The whole training process is shown in Algorithm 1, the procedure of our method requires an initial guess to $\boldsymbol{\beta}$ in the alternating optimization, where each entry of $\boldsymbol{\beta}$ is initialized with equal weights. The whole algorithm is terminated when a stopping criterion is achieved. Here a simple stopping criterion is adopted based on the variation of $\boldsymbol{\beta}$ between two consecutive iterative steps.

3   Experiments

This section first describes our implementation details and experimental setup. Then, we compare our method with the state-of-the-art methods in the literature.

3.1   Experimental Setting

Dataset To evaluate the performance of our method, we employ two widely used datasets, including TORONTO [4] and MIT [13]. The first dataset contains 120 color images with resolution of 511×681 pixels from indoor and outdoor environments. Images are presented at random to twenty human subjects for 3 s with 2 s of gray mask in between. For the second dataset, it is a larger dataset containing 1003 images (resolution from 405×1024 to 1024×1024 pixels) collected from Flicker and LabelMe datasets. There are 779 landscape and 228 portrait images. The ground truth saliency maps are generated using the eye fixation data collected from fifteen human subjects, where each subject is asked to freely view images for 3 s with 1 s delay in between.

Baselines To show the advantages of our approach, we selected 6 state-of-the-art models as baselines, including spectral residual saliency (SR [9]), attention measure (IT [12]), unified saliency (US [14]), nature statistic saliency (SUN [27]), frequency-tuned saliency (FT [1]), and co-bootstrapping saliency (CS [16]). Besides, we directly borrow three feature maps (CESC, CSC, and GC) as baselines for comprehensive comparison.

Evaluation Metrics We utilize receiver operating characteristic (ROC) curve to evaluate our system. Under this criteria, each predicted saliency map is thresholded to generate the final map. The pixels with larger saliency values than the threshold are identified as salient (positive samples), and the other pixels are considered as non-salient (negative samples) [4]. The ROC curve is plotted with the true positive rate against the false positive rate under varying threshold. After that, we also compute the area under ROC curve (AUC) score for direct comparison. As discussed in [27], however, there is always a center bias that our HVS always prefers to the center of an image. Therefore, we turn to the shuffled AUC (sAUC) score [27] as an alternative metric.

3.2   Results and Analysis

Table 1 shows the performance comparison between the proposed WLMKL method and the baseline methods in terms of the AUC and sAUC. The corresponding ROC curves are illustrated in Fig. 2. Results show that our WLMKL method outperforms

Table 1 Performance comparison of the baseline methods and our approach on two datasets in terms of AUC and sAUC. The best results are shown in bold values

	Toronto [4]		MIT [13]	
Methods	AUC	sAUC	AUC	sAUC
IT [12]	0.739	0.627	0.725	0.614
US [14]	0.815	0.670	0.804	0.658
CS [16]	0.817	0.659	0.814	0.656
FT [1]	0.534	0.447	0.515	0.422
SR [9]	0.516	0.409	0.544	0.437
SUN [27]	0.670	0.505	0.722	0.609
CESC	0.691	0.671	0.677	0.603
GC	0.816	0.690	0.808	0.676
CSC	0.811	0.694	0.816	0.670
Ours	**0.827**	**0.702**	**0.843**	**0.697**

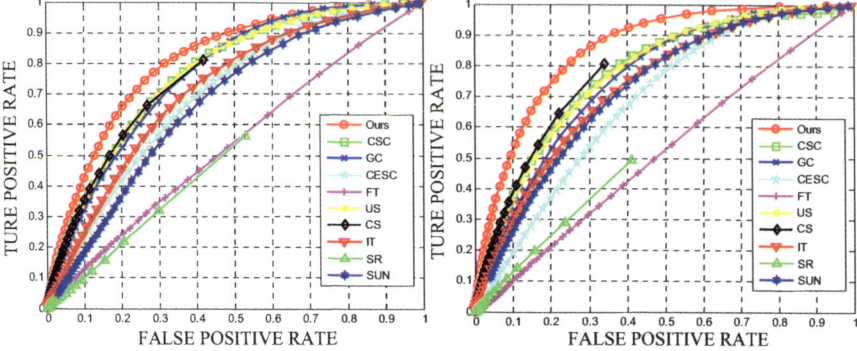

Fig. 2 ROC curve comparison between our method and other baseline approaches. From left to right are the results on the TORONTO and MIT datasets. (Best viewed in color)

the state-of-the-art approaches. From Table 1, our method averagely improves the results with 0.020 and 0.035 in terms of AUC and sAUC, and outperforms the best performing baselines with a margin of 0.031 and 0.043, respectively. We also show the promising results of each contrast feature map (CSC, GC, and CESC) over two datasets, especially the CSC almost gains higher performance than CESC, achieving comparable results with GC contrast measure. In particular, on the TORONTO dataset, CSC achieves an AUC of 0.827 and a sAUC value of 0.702, while on the MIT dataset, CSC achieves an AUC of 0.854 and a sAUC value of 0.697.

Some examples of the saliency maps produced from our WLMKL and the baseline methods are shown in Fig. 3. One can observe that WLMKL produces saliency maps more consistent with the ground truth, compared with other baselines. These results clearly demonstrate the effectiveness of WLMKL in combining the contrast feature maps to perform visual saliency detection. It is worth noting that the proposed WLMKL does not require any preprocessing such as over-segmentation, nor any assistance from the top-down priors.

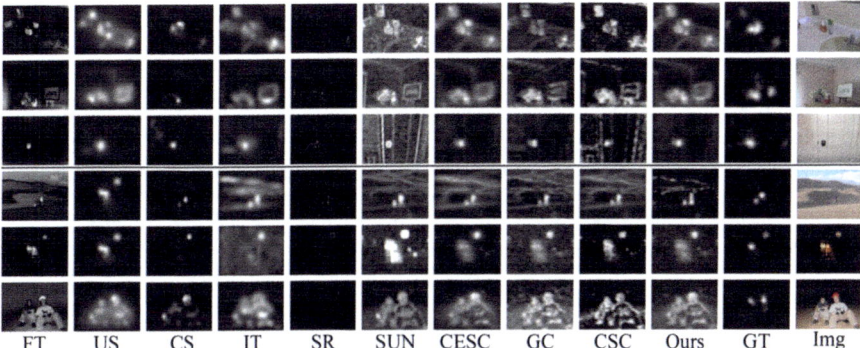

| FT | US | CS | IT | SR | SUN | CESC | GC | CSC | Ours | GT | Img |

Fig. 3 Visual comparison between our method and other baseline approaches. From top to bottom are some examples of predicted saliency maps on the TORONTO and MIT datasets. The columns from left to right, respectively, show estimated saliency maps produced by FT, US, CS, IT, SR, SUN, CESC, GC, CSC, and our methods, with corresponding ground truth and original images. (Best viewed in color)

4 Conclusion

In this paper, a WLMKL framework is proposed for visual saliency detection. WLMKL learns adaptive weights to incorporate three contrast feature maps namely, CSC, CESC, and GC, respectively. Our WLMKL model enables each contrast feature map to contribute to predict pixel saliency via preserving salient features and suppressing the non-salient features. Extensive experiments well validate the effectiveness of our framework on TORONTO and MIT benchmark datasets. In the future, we would like to explore more feature space (e.g., texture feature and edge strength) to further enhance the performance.

Acknowledgements This work was partly supported by the National Science Foundation (Grant No. IIS-1302164), the National Natural Science Foundation of China (Grant No. 61881240048, 61571240, 61501247, 61501259, 61671253), China Postdoctoral Science Foundation (Grant No. 2015M581841), Open Fund Project of Key Laboratory of Intelligent Perception and Systems for High-Dimensional Information of Ministry of Education (Nanjing University of Science and Technology) (Grant No. JYB201709, JYB201710), and Natural Science Foundation of Jiangsu Province, China (BK20160908), NUPTSF (Grant No. NY214139).

References

1. Achanta, R., Hemami, S., Estrada, F., & Süsstrunk, S. (2009). Frequency-tuned salient region detection. In *2009 IEEE Conference on Computer Vision and Pattern Recognition* (pp. 1597–1604). Piscataway: IEEE.
2. Alexe, B., Deselaers, T., & Ferrari, V. (2010) What is an object? In *2010 IEEE Computer Society conference on Computer Vision and Pattern Recognition* (pp. 73–80). Piscataway: IEEE.

3. Borji, A. (2012). Boosting bottom-up and top-down visual features for saliency estimation. In *2012 IEEE Conference on Computer Vision and Pattern Recognition* (pp. 438–445). Piscataway: IEEE.
4. Bruce, N., & Tsotsos, J. (2006). Saliency based on information maximization. In *Proceedings of the 18th International Conference on Neural Information Processing System* (pp. 155–162). Cambridge, MA: MIT Press.
5. Cheng, M., Zhang, G., Mitra, N., Huang, X., & Hu, S. (2011). Global contrast based salient region detection. In *Conference on Computer Vision and Pattern Recognition* (pp. 409–416).
6. Goferman, S., Zelnik-Manor, L., & Tal, A. (2012). Context-aware saliency detection. *IEEE Transactions on Pattern Analysis and Machine Intelligence, 34*(10), 1915–1926.
7. Gönen, M., & Alpaydın, E. (2011). Multiple kernel learning algorithms. *Journal of Machine Learning Research, 12*, 2211–2268.
8. Gopalakrishnan, V., Hu, Y., & Rajan, D. (2009). Salient region detection by modeling distributions of color and orientation. *IEEE Transactions on Multimedia, 11*(5), 892–905.
9. Hou, X., & Zhang, L. (2007). Saliency detection: A spectral residual approach. In *2007 IEEE Conference on Computer Vision and Pattern Recognition* (pp. 1–8). Piscataway: IEEE.
10. Hou, X., & Zhang, L. (2008). Dynamic visual attention: Searching for coding length increments. In *Advances in Neural Information Processing Systems* (pp. 681–688).
11. Hu, Y., Xie, X., Ma, W. Y., Chia, L. T., & Rajan, D. (2005). Salient region detection using weighted feature maps based on the human visual attention model. In *Advances in Multimedia Information Processing-PCM* (pp. 993–1000).
12. Itti, L., Koch, C., & Niebur, E. (1998). A model of saliency-based visual attention for rapid scene analysis. *IEEE Transactions on Pattern Analysis and Machine Intelligence, 20*(11), 1254–1259.
13. Judd, T., Ehinger, K., Durand, F., & Torralba, A. (2009). Learning to predict where humans look. In *2009 IEEE 12th International Conference on Computer Vision* (pp. 2106–2113). Piscataway: IEEE.
14. Kruthiventi, S. S. S., Gudisa, V., Dholakiya, J. H., & Babu, R. V.: Saliency unified: A deep architecture for simultaneous eye fixation prediction and salient object segmentation. In *2016 IEEE Conference on Computer Vision and Pattern Recognition (CVPR)* (pp. 5781–5790). Piscataway: IEEE.
15. Li, J., Li, X., Yang, B., & Sun, X. (2015). Segmentation-based image copy-move forgery detection scheme. *IEEE Transactions on Information Forensics and Security, 10*(3), 507–518.
16. Lu, H., Zhang, X., Qi, J., Tong, N., Ruan, X., & Yang, M. H. (2017). Co-bootstrapping saliency. *IEEE Transactions on Image Processing, 26*(1), 414–425.
17. Ma, Q., & Zhang, L. (2008). Image quality assessment with visual attention. In *2008 19th International Conference on Pattern Recognition* (pp. 1–4). Piscataway: IEEE.
18. Mairal, J., Bach, F., Ponce, J., & Sapiro, G. (2010). Online learning for matrix factorization and sparse coding. *Journal of Machine Learning Research, 11*, 19–60 (2010)
19. Marchesotti, L., Cifarelli, C., G., & Gabriela, C. (2009). A framework for visual saliency detection with applications to image thumbnailing. In *2009 IEEE 12th International Conference on Computer Vision* (pp. 2232–2239).
20. Olshausen, B. A., & Field, D. J. (1996). Emergence of simple-cell receptive field properties by learning a sparse code for natural images. *Nature, 381*(6583), 607–609.
21. Pan, Z., Zhang, Y., & Kwong, S. (2015). Efficient motion and disparity estimation optimization for low complexity multiview video coding. *IEEE Transactions on Broadcasting, 61*(2), 166–176.
22. Shen, X., & Wu, Y. (2012). A unified approach to salient object detection via low rank matrix recovery. In *2012 IEEE Conference on Computer Vision and Pattern Recognition* (pp. 853–860). Piscataway: IEEE.
23. Simoncelli, E. P., & Olshausen, B. A. (2001). Natural image statistics and neural representation. *Annual Review of Neuroscience, 24*(1), 1193–1216.
24. Treisman, A. M., & Gelade, G. (1980). A feature-integration theory of attention. *Cognitive Psychology, 12*(1), 97–136.

25. Vapnik, V. (1993). *The Nature of Statistical Learning Theory*. Berlin: Springer.
26. Yu, J. G., Xia, G. S., Gao, C., & Samal, A. (2016). A computational model for object-based visual saliency: Spreading attention along gestalt cues. *IEEE Transactions on Multimedia, 18*(2), 273–286.
27. Zhang, L., Tong, M. H., Marks, T. K., Shan, H., & Cottrell, G. W. (2008). SUN: A Bayesian framework for saliency using natural statistics. *Journal of Vision, 8*(7), 32.
28. Zhou, Q. (2014). Object-based attention: saliency detection using contrast via background prototypes. *Electronics Letters, 50*(14), 997–999.
29. Zhou, Q., Li, N., Yang, Y., Chen, P., & Liu, W. (2012). Corner-surround contrast for saliency detection. In *Proceedings of the 21st International Conference on Pattern Recognition (ICPR2012)* (pp. 1423–1426). Piscataway: IEEE.
30. Zhou, Q., Chen, J., Ren, S., Zhou, Y., Chen, J., & Liu, W. (2013). On contrast combinations for visual saliency detection. In *2013 IEEE International Conference on Image Processing* (pp. 2665–2669). Piscataway: IEEE.
31. Zhou, Q., Cai, S., Zhu, S., & Zheng, B. (2014). Salient object detection using window mask transferring with multi-layer background contrast. In *Asian Conference on Computer Vision* (pp. 221–235). Cham: Springer.

Index

© Springer Nature Switzerland AG 2020
H. Lu, L. Yujie (eds.), *2nd EAI International Conference on Robotic Sensor
Networks*, EAI/Springer Innovations in Communication and Computing,
https://doi.org/10.1007/978-3-030-17763-8

Printed by Printforce, the Netherlands